KU-724-584

Level 2

CARPENTRY & JOINERY

3rd Edition

NVQ/SVQ and CAA Diploma

carillion

www.pearsonschoolsandfecolleges.co.uk

✓ Free online support
✓ Useful weblinks
✓ 24 hour online ordering

0845 630 44 44

Heinemann

Part of Pearson

Heinemann is an imprint of Pearson Education Limited, a company incorporated in England and Wales, having its registered office at Edinburgh Gate, Harlow, Essex, CM20 2JE. Registered company number: 872828

www.pearsonschoolsandfecolleges.co.uk

Heinemann is a registered trademark of Pearson Education Limited

Text © Carillion Construction Ltd 2010

First published 2010

14 13 12 11
10 9 8 7 6 5 4 3 2

British Library Cataloguing in Publication Data is available from the British Library on request.

ISBN 978 0 435027 04 9

Copyright notice

Typeset by Tek-Art
Original illustrations © Pearson Education Ltd
Illustrated by Oxford Designers and Illustrators
Cover design by Wooden Ark
Printed in Malaysia, CTP-KHL

Acknowledgements

Every effort has been made to contact copyright holders of material reproduced in this book. Any omissions will be rectified in subsequent printings if notice is given to the publishers.

Websites

The websites used in this book were correct and up-to-date at the time of publication. It is essential for tutors to preview each website before using it in class so as to ensure that the URL is still accurate, relevant and appropriate. We suggest that tutors bookmark useful websites and consider enabling students to access them through the school/college intranet.

The information and activities in this book have been prepared according to the standards reasonably to be expected of a competent trainer in the relevant subject matter. However, you should be aware that errors and omissions can be made and that different employers may adopt different standards and practices over time. Before doing any practical activity, you should always carry out your own Risk Assessment and make your own enquires and investigations into appropriate standards and practices to be observed.

Acknowledgements

Carillion would like to thank the following people for their contribution to this book: Kevin Jarvis and John Harvie McLaughlin.

Pearson Education Limited would like to thank the following people for providing technical feedback: Graeme Coventry from Basingstoke College of Technology; Rob Harrison from Stoke College and Jim Neil and Gerry McGowan from North Glasgow College.

Photo acknowledgements
The author and publisher would like to thank the following individuals and organisations for permission to reproduce photographs:

(Key: b-bottom; c-centre; L-left; r-right; t-top)

Alamy: Adrian Sherratt 50; Alex Segre 197r; Altrendo 285; Clivestock 188; Daniel Goodchild 186/9.45; David J. Green 33; Don John Red 197L; Eric Nathan 91; Geoff du Feu 49; Image Source Pink 14; Imagebroker 87t; Janice Hazeldine 169; Jon Parker Lee 186c; Mike Rinnan 272; Niall McDiarmid 160; Peter Jordan 309/34.12; British Gypsum: 278 (all); Construction Photography: Chipboard Flooring 106; Darren Holden 321; David R. Frazier 155r; David Stewart-Smith 105; DIY Photolibrary 104t, 155L; Ken Price 26b; Paul McMullin 155c; Sally-Ann Norman 154t; Corbis: Creasource 51; Cusp 181; CSCS: 9; Getty Images: Photodisc 15, 43; Red Cover 335; Stone 116; iStockPhoto: Ales Veluscek 133 (whitewood); Bill Noll 133 (cedar, ash, oak), 134 (African, Maple, Sapele, Teak), 135; Dave White 32 (fir); Dimitriy Pochitalin 134 (beech); Flashon Studios 215; George Peters 214/9.136; Ivan Vasilev 132 (larch); Nancy Nehring 133 (redwood); Tibor Nagy 133 (mahogany); Pearson Education: Chris Honeywell 36/1.17; David Sanderson 37r, 38; Jules Selmes 13, 104b, 157 (all), 179, 180 (all), 182 (9.29-9.32), 279; Photographers Direct: John Gavin 346; Martyn F. Chillmaid 206/9.117; Will Burwell 309/34.13; Science Photo Library: Garry Watson 17L; Shutterstock: AG Photography 261; Alex Kosev 36/1.16; Andrey Bayda 17c; Chepko Danil Vitalevich 303; Colour 87b; Dainis Derics 206/9.113; Diego Cervo 1; Edd Westmacott 36/1.21, 310/34.16; Frances A. Miller 26t; IOFoto 36/1.19; Jiri Hera 205/9.111; Joachim Wendler 308; Josef Bosak 205/9.108; Martina Orlich 132 (pine); Michael Shake 36/1.18; Mitch Aunger 273; Nikolay Okhtin 83; Nito 185/9.40; Quayside 187/9.51; Rob Byron 36/1.20; Sheldunov Andrew 219; StillFX 37L; Sue Smith 228; Yobidaba 17r; Yuri Arcurs 14 (inset); Toolbank: 415

Cover image: Alamy Images: Image Source Black

All other images © Pearson Education Ltd / Gareth Boden

Picture Research by: Chrissie Martin

Contents

Introduction

Welcome to NVQ/SVQ CAA Diploma Level 2 Carpentry & Joinery!

Carpentry & Joinery combines many different practical and visual skills with knowledge of specialised materials and techniques. This book will introduce you to the construction trade and in particular the knowledge and skills needed for interpreting site documents, basic woodworking joints, and safe storage and use of a selection of hand and power tools.

About this book

This book has been produced to help you build a sound knowledge and understanding of all aspects of the Diploma and NVQ requirements associated with Carpentry & Joinery.

The information in this book covers the information you will need to attain your Level 2 qualification in Carpentry & Joinery. Each chapter of the book relates to a particular unit of the CAA Diploma and provides the information needed to form the required knowledge and understanding of that area. The book is also designed to support those undertaking the NVQ at Level 2.

This book has been written based on a concept used with Carillion Training Centres for many years. The concept is about providing learners with the necessary information they need to support their studies and at the same time ensuring it is presented in a style which is both manageable and relevant.

This book will also be a useful reference tool for you in your professional life once you have gained your qualifications and are a practicing carpenter or joiner.

This introduction will introduce the construction industry and the qualifications you can find in it, alongside the qualifications available.

Qualifications for the construction industry

There are many ways of entering the construction industry, but the most common method is as an apprentice.

Apprenticeships

You can become an apprentice by being employed:

- directly by a construction company who will send you to college
- by a training provider, such as Carillion, which combines construction training with practical work experience.

Construction Skills is the national training organisation for construction in the UK and is responsible for setting training standards.

The framework of an apprenticeship is based around an NVQ (or SVQ in Scotland). These qualifications are developed and approved by industry experts and will measure your practical skills and job knowledge on-site.

You will also need to achieve:

- a technical certificate
- the Construction Skills health and safety test
- the appropriate level of Functional skills assessment
- an Employees Rights and Responsibilities briefing.

You will also need to achieve the right qualifications to get on a construction site, including qualifying for the CSCS card scheme.

CAA Diploma

The Construction Awards Alliance (CAA) Diploma was launched on 1 August 2008 to replace Construction Awards. They aim to make you:

- more skilled and knowledgeable
- more confident with moving across projects, contracts and employers

The CAA Diploma is a common testing strategy with knowledge tests for each unit, a practical assignment and the GOLA (Global Online Assessment) test.

The CAA Diploma meets the requirements of the new Qualifications and Credit Framework (QCF) which bases a qualification the number of credits (with ten learning hours gaining one credit):

- Award (1 to 12 credits)
- Certificate (13 to 36 credits)
- Diploma (37+ credits)

As part of the CAA Diploma you will gain the skills needed for the NVQ as well as the Functional skills knowledge you will need to complete your qualification.

National Vocational Qualifications (NVQs)

NVQs are available to anyone, with no restrictions on age, length or type of training, although learners below a certain age can only perform certain tasks. There are different levels of NVQ (for example 1, 2, 3), which in turn are broken down into units of competence. NVQs are not like traditional examinations in which someone sits an exam paper. An NVQ is a 'doing' qualification, which means it lets the industry know that you have the knowledge, skills and ability to actually 'do' something.

NVQs are made up of both mandatory and optional units and the number of units that you need to complete for an NVQ depends on the level and the occupation.

NVQs are assessed in the workplace, and several types of evidence are used:

- Witness testimony provided by individuals who have first-hand knowledge of your work and performance relating to the NVQ.
- Your performance can be observed a number of times in the workplace.
- Historical evidence means that you can use evidence from past achievements or experience, if it is directly related to the NVQ.
- Assignments or projects can be used to assess your knowledge and understanding.
- Photographic evidence showing you performing various tasks in the workplace can be used, providing it is authenticated by your supervisor.

Introduction

Introduction

Functional skills

This feature is designed to support you with your functional skills, by identifying opportunities in your work where you will be able to practice your functional skills.

Features of this book

This book has been fully illustrated with artworks and photographs. These will help to give you more information about a concept or a procedure, as well as helping you to follow a step-by-step procedure or identify a tool or material.

This book also contains a number of different features to help your learning and development.

Remember

This highlights key facts or concepts, sometimes from earlier in the text, to remind you of important things you will need to think about.

Did you know?

This feature gives you interesting facts about the building trade

Safety tip

This feature gives you guidance for working safely on the tasks in this book

Find out

These are short activities and research opportunities, designed to help you gain further information about, and understanding of, a topic area.

Key term

These are new or difficult words. They are picked out in **bold** in the text and then defined in the margin.

Working life

This feature gives you a chance to read about and debate a real life work scenario or problem. Why has the situation occurred? What would you do?

FAQ

These are frequently asked questions appearing at the end of each unit to answer your questions with informative answers from the experts

Check it out

A series of questions at the end of each unit to check your understanding. Some of these questions may support the collecting of evidence for the NVQ.

Getting ready for assessment

This feature provides guidance for preparing for the practical assessment. It will give you advice on using the theory you have learnt about in a practical way.

Knowledge check

This is a series of multiple choice questions in the style of the GOLA end of unit tests at the end of each unit.

UNIT 1001

Safe working practices in construction

Health and safety is a vital part of all construction work. All work should be completed in a way that is safe not only for the individual worker, but also for the other workers on the site, people nearby and the final users of the building.

Every year in the construction industry over 100 people are killed and many more are seriously injured as a result of the work that they do. There are thousands more who suffer from work-related health problems, such as dermatitis, asbestosis, industrial asthma, vibration white finger (see pages 15–18) and deafness. Therefore, learning as much as you can about health and safety is very important.

This unit also supports NVQ units VR01 Conform to General Workplace Safety and VR 03 Move and Handle Resources.

This unit contains material that supports TAP Unit 1 Erect and Dismantle Working Platforms. It also contains material that supports the delivery of the five generic units.

This unit will cover the following learning outcomes:

- Health and safety regulations – roles and responsibilities
- Accident/first aid/emergency procedures and reporting
- Hazards on construction sites
- Health and hygiene
- Safe handling of materials and equipment
- Basic working platforms
- Working with electricity
- Use of appropriate personal protective equipment (PPE)
- Fire and emergency procedures
- Safety signs and notices.

Key terms

Legislation – a law or set of laws passed by Parliament, often called an Act

Hazardous – something or a situation that is dangerous or unsafe

Employer – the person or company you work for

Employee – the worker

Key terms

Proactive – acting in advance, before something happens (such as an accident)

Reactive – acting after something happens, in response to it

Key terms

Subcontractor – workers who have been hired by the main contractor to carry out works, usually specialist works, e.g. a general builder may hire a plumber as a subcontractor as none of their staff can do plumbing work

Supplier – a company that supplies goods, materials or services

K1. Health and safety regulations

While at work, whatever your location or the type of work you are doing, there is important **legislation** you must comply with. Health and safety legislation is there not just to protect you – it also states what you must and must not do to ensure that no workers are placed in a situation **hazardous** to themselves or others.

There are hundreds of Acts covering all manner of work from hairdressing to construction. Each Act states the duties of the **employer** and **employee** – you should be aware of both. If an employer or employee does something they shouldn't – or doesn't do something they should – they can end up in court and be fined or even imprisoned.

Approved code of practice, guidance notes and safety policies

As well as Acts, there are two types of codes of practice and guidance notes: those produced by the Health and Safety Executive (HSE; see page 4), and those created by companies themselves. Most large construction companies – and many smaller ones – have their own guidance notes, which go further than health and safety law. For example, the law states that everyone must wear safety boots in a hazardous area, but a company's code may state that everyone must wear safety boots at all times. This is called taking a **proactive** approach, rather than a **reactive** one.

Health and safety legislation you need to be aware of

One phrase that often comes up in the legislation is 'so far as is reasonably practicable'. This means that health and safety must be adhered to at all times, but must take a common sense, practical approach.

The Health and Safety at Work Act 1974 (HASAW)

HASAW applies to all types and places of work and to employers, employees, self-employed people, **subcontractors** and even **suppliers**. The Act is there to protect not only the people at work, but also the general public, who may be affected in some way by the work that has been or will be carried out.

The main objectives of the Health and Safety at Work Act are to:

- ensure the health, safety and welfare of all persons at work

- protect the general public from all work activities
- control the use, handling, storage and transportation of explosives and highly **flammable** substances
- control the release of noxious or offensive substances into the atmosphere.

To ensure that these objectives are met there are duties for all employers, employees and suppliers.

The employer's duties

Employers must:

- provide safe **access** and **egress** to and within the work area
- provide a safe place to work
- provide and maintain plant and machinery that is safe and without risks to health
- provide information, instruction, training and supervision to ensure the health and safety at work of all employees
- ensure safety and the absence of risks to health in connection with the handling, storage and transportation of articles and substances
- have a written safety policy that must be revised and updated regularly, and ensure all employees are aware of it
- involve trade union safety representatives, where appointed, in all matters relating to health and safety
- carry out risk assessments (see page 14) and provide supervision where necessary
- provide and not charge for personal protective equipment (**PPE**).

The employee's duties

Employees must:

- take reasonable care for their own health and safety
- take reasonable care for the health and safety of anyone who may be affected by their acts or **omissions**
- co-operate with their employer or any other person to ensure legal **obligations** are met
- not misuse or interfere with anything provided for their health and safety
- report hazards and accidents (see page 10–11)
- use any equipment and safeguards provided by their employer.

Key terms

Flammable – something that is easily lit and burns rapidly

Access – entrance, a way in

Egress – exit, a way out

PPE – personal protective equipment such as gloves, a safety harness or goggles

Remember

Employers can't charge their employees for anything that has been done or provided for them to ensure that legal requirements on health and safety are met. Self-employed people and subcontractors have the same duties as employees. If they have employees of their own, they must also obey the duties set down for employers

Key terms

Omission – something that has not been done or has been missed out

Obligation –something you have a duty or a responsibility to do

Unit 1001 Safe working practices in construction

3

The supplier's duties

Persons designing, manufacturing, importing or supplying articles or substances for use at work must ensure that:

- articles are designed and constructed so that they will be safe and without risk to health at all times while they are being used or constructed
- substances will be safe and without risk to health at all times when being used, handled, transported and stored
- tests on articles and substances are carried out as necessary
- adequate information is provided about the use, handling, transporting and storing of articles or substances.

Health and Safety Executive (HSE)

HASAW, like most of the other Acts mentioned, is enforced by the HSE.

The HSE is the government body responsible for the encouragement, regulation and enforcement of health, safety and welfare in the workplace in the UK. It also has responsibility for research into occupational risks in England, Wales and Scotland. In Northern Ireland the responsibility lies with the Health and Safety Executive for Northern Ireland.

The HSE's duties are to:

- assist and encourage anyone who has any dealings with the objectives of HASAW
- produce and encourage research, publication, training and information on health and safety at work
- ensure that employers, employees, suppliers and other people are provided with an information and advisory service, and are kept informed and advised on any health and safety matters
- propose regulations
- enforce HASAW.

To aid in theses duties the HSE has several resources, including a laboratory used for, among other things, research, development and **forensic investigation** into the causes of accidents. The enforcement of HASAW is usually delegated to local government bodies such as county or district councils.

An enforcing authority may appoint **inspectors**, who, under the authority, have the power to:

- enter any premises which she or he has reason to believe it is necessary to enter to enforce the Act, at any reasonable time, or in a dangerous situation

Did you know?

The HSE now also includes the Health and Safety Commission (HSC) which was merged with it in 2008

Did you know?

Local government bodies can be **enforcing authorities** for several workplaces, including offices, shops, retail and wholesale distribution, hotel and catering establishments, petrol filling stations, residential care homes and the leisure industry

Key terms

Forensic investigation – a branch of science that looks at how things happen

Enforcing authorities – an organisation or people who have the authority to enforce certain laws or Acts, as well as providing guidance or advice

Inspectors – someone who is appointed or employed to inspect/examine something in order to judge its quality or compliance with any laws

- bring a police constable if there is reasonable cause to fear any serious obstruction in carrying out their duty
- bring any other person authorised by the enforcing authority, and any equipment or materials required
- examine and investigate any circumstance that is necessary for the purpose of enforcing the Act
- give orders that the premises, any part of them or anything therein, shall be left undisturbed for so long as is needed for the purpose of any examination or investigation
- take measurements, photographs and make any recordings considered necessary for the purpose of examination or investigation
- take samples of any articles or substances found and of the atmosphere in or in the vicinity of the premises
- have an article or substance which appears to be a danger to health or safety, dismantled, tested or even destroyed if necessary
- take possession of a hazardous article and detain it for so long as is necessary in order to examine it and ensure that it is not tampered with and that it is available for use as evidence in any **prosecution**
- interview any person believed to have information, ask any questions the inspector thinks fit to ask and ensure all statements are signed as a declaration of the truth of the answers
- require the production of, inspect and take copies of, any entry in any book or document which it is necessary for the purposes of any examination or investigation
- utilise any other power which is necessary to enforce the Act.

> **Key term**
>
> **Prosecution** – accusing someone of committing a crime, which usually results in the accused being taken to court and, if found guilty, being punished

Contacting the HSE

Employers, self-employed people or someone in control of work premises have legal duties to record and report to the HSE some work-related accidents. Incidents that must be reported are:

- death, major injury or disease
- **dangerous occurrence** – an event that may not have caused injury, but could have done so
- **over three day injury** – an injury at work that results in the worker being away from work for more than three consecutive days.

Construction (Design and Management) Regulations 2007

The Construction (Design and Management) Regulations 2007, often referred to as CDM, are important regulations in the construction industry. They were introduced by the HSE's Construction Division. The regulations deal mainly with the construction industry and aim to improve safety.

The duties for employers under the regulations are to:

- plan, manage and monitor their own work and that of workers
- check competence of all their appointees and workers
- train their employees
- provide information to their workers
- comply with the specific requirements in Part 4 of the Regulations, which deals with lighting, excavations, traffic routes, etc.
- ensure there are adequate welfare facilities for their workers.

The duties for employees are to:

- check their own competence
- co-operate with others and co-ordinate work so as to ensure the health and safety of construction workers and others who may be affected by the work
- report obvious risks.

> **Did you know?**
>
> On large projects, a person is appointed as the CDM co-ordinator. This person has overall responsibility for compliance with CDM. There is a general expectation by the HSE that all parties involved in a project will co-operate and co-ordinate with each other

Provision and Use of Work Equipment Regulations 1998 (PUWER)

These regulations cover all new or existing work equipment – leased, hired or second-hand. They apply in most working environments where the HASAW applies, including all industrial, offshore and service operations. PUWER covers the starting, stopping, regular use, transport, repair, modification, servicing and cleaning of equipment.

'Work equipment' includes any machinery, appliance, apparatus or tool, and any assembly of components that are used in non-domestic premises.

The general duties of the Act require equipment to be:

- suitable for its intended purpose and to be used only in suitable conditions
- maintained in an efficient state and maintenance records kept
- used, repaired and maintained only by a suitably trained person when that equipment poses a particular risk
- able to be isolated from all its sources of energy

> **Did you know?**
>
> Dumper trucks, circular saws, ladders, overhead projectors and chisels are all covered by PUWER, but substances, private cars and structural items are not

- constructed or adapted to ensure that maintenance can be carried out without risks to health and safety
- fitted with warnings or warning devices as appropriate.

In addition, the Act requires:

- all those who use, supervise or manage work equipment to be suitably trained
- access to any dangerous parts of the machinery to be prevented or controlled
- injury to be prevented from any work equipment that may have a very high or low temperature
- suitable controls to be provided for starting and stopping the work equipment
- suitable emergency stopping systems and braking systems to be fitted to ensure the work equipment is brought to a safe condition as soon as reasonably practicable
- suitable and sufficient lighting to be provided for operating the work equipment.

Other pieces of legislation

Legislation	Content
The Reporting of Injuries, Diseases and Dangerous Occurrences Regulations 1995 (RIDDOR)	Employers have a duty to report accidents, diseases or dangerous occurrences. The HSE uses this information to identify where and how risk arises and to investigate serious accidents.
Control of Substances Hazardous to Health Regulations 2002 (COSHH)	State how employees and employers should work with, handle, store, transport and dispose of potentially hazardous substances. This includes substances used and generated during work (e.g. paints or dust), naturally occurring substances (e.g. sand) and biological elements (e.g. bacteria).
The Control of Noise at Work Regulations 2005	Protect employees from consequences of exposure to noise. These state employers must: • assess the risks to the employee and make sure legal limits are not exceeded • take action to reduce noise exposure and provide hearing protection • provide information, instruction and training.
The Electricity at Work Regulations 1989	Cover any work involving electricity or electrical equipment. Employers must keep electrical systems safe and regularly maintained and do everything to reduce the risk of employers coming into contact with live electrical currents.
The Manual Handling Operations Regulations 1992	Cover all work activities involving a person lifting. Where possible, manual handling should be avoided, but where unavoidable a risk assessment must be carried out, focusing on the load, the individual, the task and the environment.

Unit 1001

Safe working practices in construction

Legislation	Content
The Personal Protective Equipment at Work Regulations 1992 (PPER)	Cover all types of PPE. PPE must be checked prior to use by a trained and competent person, in line with manufacturer's instructions. PPE must be provided by the employer free of charge with a suitable and secure storage place. Employers must ensure employees know the risks PPE will avoid, its purpose, how to maintain it and its limitations. Employees must ensure they are trained to use PPE, they use it in line with instructions, return it to storage after use and report any loss or defect.
The Work at Height Regulations 2005	Ensure employers do all they can to reduce the risk of injury or death from working at height. Employers must avoid working at height where possible and use equipment/safeguards that prevent or minimise the danger of falls. Employees must follow any training, report hazards and use any safety equipment provided.

Remember

Legislation can change or be updated. New legislation can be created as well – this could even supersede all pieces of legislation

Health and safety is a large and varied subject that changes regularly. The introduction of new regulations or updates to current legislation means it's often hard to remember or keep up to date. Your tutor will be able to give you information on current legislation.

Any future employers should also keep you updated on any changes to legislation that will affect you. There are also other sources of information that can be accessed to keep you informed. The main sources of Health and Safety Information are shown in the table below.

The Health and Safety Executive	Wide range of information ranging from actual legislation to helpful guides. The website has videos, leaflets and documents available for free download, with specific sections dedicated to different industries. The construction website is www.hse.gov.uk/construction.
Construction Skills	As well as advice on qualifications, they also offer advice on health and safety matters and on sitting the CSCS health and safety test. The website address is www.cskills.org.
Royal Society for the Prevention of Accidents (RoSPA)	Provides information, advice, resources and training and are actively involved in the promotion of safety and accident prevention. The website address is www.rospa.com.
Royal Society for the Promotion of Health (RSPH)	An independent organisation with the goal of promoting and protection of health. Uses advocacy, mediation, knowledge and practice to advise on policy development. Also provides education, training, research, communicates information and provides certification for products, training centres and processes. The website address is www.rsph.org/en/health-promotion.

Site inductions

Site inductions are the process that an individual undergoes in order to accelerate their awareness of the potential health and safety hazards and risks they may face in their working environment; this excludes job related skills training.

Different site inductions will include different topics depending on the work that is being carried out. The basic information inductions should contain is:

- the scope of operations carried out at the site, project, etc.
- the activities that have health and safety hazards and risks
- the control measures, emergency arrangements and welfare arrangements in place
- the local organisation and management structure
- the consultation procedures and resources for health and safety advice
- the process for reporting near misses.

Inductions are also vital for informing all people working on the site of amenities, restricted areas, dress code (PPE) and even evacuation procedures. Inductions must be carried out by a competent person. Records of all inductions must be kept to ensure that all workers have received an induction. Some sites will even hand out cards to those who have been inducted and people without cards will not be admitted to the site.

Toolbox talks

Toolbox talks are a vital tool used by management, supervisors and employees to deliver basic training and/or to inform all workers of any updates to policy, hazardous activities/areas or other information. They should be delivered by a competent person and a record of all attendees should be kept.

Toolbox talks should be relevant to the people they are being delivered to. The topics can vary from being informative, e.g. letting everyone know a reclassification of a PPE area, to basic training on the use of a certain tool.

Construction Skills Certification Scheme (CSCS)

The Construction Skills certification scheme was introduced to help improve the quality of work and

> **Remember**
>
> A site induction must take place BEFORE you start work on that site

> **Remember**
>
> Visitors to the site who may not be doing any work should still receive an induction of sorts as they also need to be aware of amenities, restricted areas and procedures, etc.

> **Did you know?**
>
> Toolbox talks don't need to be formal meetings but can be held in a canteen at break time. However, a list of all attendees must be kept to ensure that everyone who needs to attend the talk does so

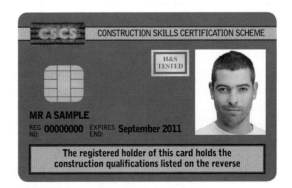

Figure 1.1 CSCS card

to reduce accidents. It requires all workers to obtain a CSCS card before they are allowed to carry out work on a building site. There are various levels of cards which indicate your competence and skills background. This ensures that only skilled and safe tradespeople can carry out the required work on site.

To get a CSCS card all applicants must sit a health and safety test. The aim of the test is to examine knowledge across a wide range of topics in order to improve safety and productivity on site. It is usually taken as a PC-based touch screen test, either at a mobile testing unit or an accredited test centre. The type of card being applied for will determine the level of test that needs to be taken.

As a trainee, once you pass the health and safety test you will qualify for a trainee card and once you have achieved a Level 2 qualification you can then upgrade your card to an experienced worker card. Achieving a Level 3 qualification allows you to apply for a gold card. People who make regular visits to site can apply for a visitor card.

K2. Accident, first aid and emergency procedures and reporting

Major types of emergency

There are several major types of emergencies that could occur on site. These include not only accidents, but also fires, security alerts and bomb alerts. At your site induction, it should be made perfectly clear to you what you should do in the event of an emergency.

Reporting accidents

All accidents need to be reported and recorded in the accident book and the injured person must report to a trained first aider in order to receive treatment. Serious accidents must be reported under the Reporting of Injuries, Diseases and Dangerous Occurrences Regulations 1995 (RIDDOR).

The nature and seriousness of the accident will decide who it needs to be reported to. There are several types of documentation used to record accidents and emergencies.

The accident book

The accident book is completed by the person who had the accident or, if this is not possible, someone who is representing the injured person.

Safety tip

You should also be aware of any sirens or warning noises that accompany each and every type of emergency such as bomb scares or fire alarms. Some sites may have different variations of sirens or emergency procedures, so it is vital that you pay attention and listen to all instructions. If you are unsure always ask

Remember

Most accidents are caused by human error, which means someone has done something they shouldn't have done or, just as importantly, not done something they should have done. Accidents often happen when someone is hurrying, not paying enough attention to what they are doing or when they have not received the correct training

Remember

An accident that falls under RIDDOR should be reported by the safety officer or site manager and can be reported to the HSE by telephone (08453009923) or via the RIDDOR website (www.riddor.gov.uk)

Relevant authorised persons	What to do
First aiders	All accidents must be reported to a first aider. If you are unsure who they are or cannot contact them, report it to your supervisor.
Supervisors	Must be informed so they can inform the first aider and their manager, and stop the work if necessary to prevent further accidents.
Safety officers	Will be alerted by your supervisor or site manager and will asses the area to check it is safe, investigate the cause of the accident and prepare a report for HSE (if needed).
HSE	Must be reported to immediately if the accident results in death or major injury and followed up by a written report within ten days. This is done on form F2508.
Managers	Should be informed by the supervisor or safety officer as they may need to report to head office. They will also be tasked with contacting the HSE.
Emergency services	Should be contacted as soon as possible. Usually first aiders contact ambulances and supervisors contact the fire brigade. If in doubt call the emergency services yourself.

Some basic details about the accident will have to be documented in the accident book, including:

- who was involved
- what happened and where
- the date and time of the accident
- any witnesses to the accident
- the address of the injured person
- what PPE was being worn
- what first aid treatment was given.

Major and minor accidents

Often an accident will result in an injury which may be minor (e.g. a cut or a bruise) or possibly major (e.g. loss of a limb). Accidents can also be fatal.

Near misses

As well as reporting accidents, 'near misses' must also be reported. A 'near miss' is when an accident nearly happened but did not actually occur. Reporting near misses might identify a problem and can prevent accidents from happening in the future. This allows a company to be proactive rather than reactive.

Work related injuries in the construction industry

Construction has the largest number of fatal injuries of all the main industry groups. In 2007–2008 there were 72 fatal injuries. This gave a rate of 3.4 people injured per 100 000 workers. The rate of fatal injuries in construction over the past decade has shown a downward trend, although the rate has shown little change in the most recent years.

- From 1999–2000 to 2006–2007 the rate of reported major injuries in construction fell. It is unclear whether the rise in 2007–2008 means an end to this trend. Despite this falling trend, the rate of major injury in construction is the highest of any main industry group (599.2 per 100 000 employees in 2007–2008).

- Compared to other industries, a higher proportion of reported injuries were caused by falls from height, falling objects, and contact with moving machinery.

- The THOR-GP surveillance scheme data (2006–2007), indicates a higher rate of work-related illness in construction than across all industries. The rate of self-reported work-related ill health in construction is similar to other industries.

The cost of accidents

As well as the tragedy, pain and suffering that accidents cause, they can also have a negative financial and business impact.

Remember

Companies that have a lot of accidents will have a poor company image for health and safety and will find it increasingly difficult to gain future contracts. Unsafe companies with many accidents will also see injured people claiming against their insurance which will see their premiums rise. This will eventually make them uninsurable meaning they will not get any work

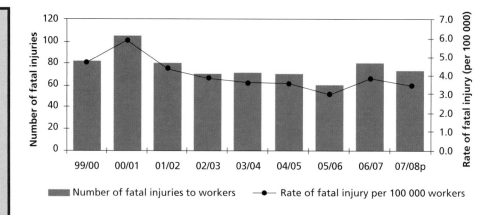

Figure 1.2 Number and rate of fatal injury to workers, 1999–2000 to 2007–2008

Small accidents will affect profits as sick pay may need to be paid. Production may also slow down or stop if the injured person is a specialist. Replacement or temporary workers may need to be used to keep the job going. More serious accidents will see the financial loss rise as the injured person will be off work for longer. This can cause jobs to fall seriously behind and, in extreme cases, may even cause the contractor to lose the job and possibly have to close the site.

First Aid

If there are more than five people on a site, a qualified first aider must be present at all times. On large building sites there must be several first aiders. A good first aid box should have plasters, bandages, antiseptic wipes, latex gloves, eye patches, slings, wound dressings and safety pins. Other equipment, such as eye wash stations, must also be available if the work being carried out requires it.

Actions for an unsafe area

On discovering an accident the first thing to do is to ensure that the victim is in no further danger. This will require you to do tasks such as switching off the electricity supply. Tasks like this must only be done if there is no danger to yourself. You should then contact the first aider or emergency services for help.

K3. Hazards on construction sites

A major part of health and safety at work is being able to identify hazards and to help prevent them in the first place, therefore avoiding the risk of injury.

Hazards on the building site

The building industry can be a very dangerous place to work and there are certain hazards that all workers need to be aware of. Some of these common hazards are covered later in this unit: falling from height (page 26–33), electrocution (page 33–34) and fires (page 38–39).

Tripping

The main cause of tripping is poor housekeeping. Whether working on scaffolding or on ground level, an untidy workplace is an accident waiting to happen. All workplaces should be kept tidy and free of debris. Not only will this prevent trip hazards, but it will also prevent costly clean-up operations at the end of the job and will promote a good professional image.

Remember

Health and safety is everyone's duty. If you receive first aid treatment and notice that there are only two plasters left, you should report it to your line manager

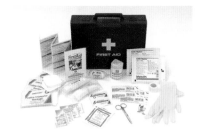

Figure 1.3 A first aid box provides the supplies to deal with minor injuries

Safety tip

Turning off the electricity is just one possible example. There will be specific safety issues for individual jobs the injured individual may have been working on. However, you should always make sure the area is safe before you continue – otherwise you could become a casualty as well

Did you know?

Housekeeping is the simple term used for cleaning up after yourself and ensuring your work area is clear and tidy. Good housekeeping is vital on a construction site as an unclean work area is dangerous. Correct storage is a big part of housekeeping and is covered on pages 22–25

Figure 1.4 An untidy work site can present many trip hazards

Remember

When a hazardous substance is being used a COSHH or risk assessment will have been made, and it should include a plan for dealing with a spillage

Key term

Carry out a risk assessment – measure the dangers of an activity against the likelihood of accidents taking place

Functional skills

When completing risk assessments you will be practising the following functional skills: FE 1.3.1 – 1.3.5: Write clearly with a level of detail to suit the purpose.

Chemical spills

Chemical spillages can range from minor inconvenience to major disaster. Most spillages are small and create minimal or no risk. If the material involved is not hazardous, it can simply be cleaned up. However, some spills may involve a hazardous material, so it is important to know what to do before the spillage happens so that remedial action can be prompt and harmful effects minimised.

Burns

Burns can occur not only from the obvious source of fire and heat but also from materials containing chemicals such as cement or painter's solvents. Electricity can also cause burns. It is vital when working with materials that you are aware of the hazards they may present and take the necessary precautions.

Risk assessments

You will have noticed that most of the legislation we have looked at requires risk assessments to be carried out. The Management of Health and Safety at Work Regulations 1999 require every employer to make suitable and sufficient assessment of:

- the risks to the health and safety of their employees to which the employees are exposed while at work
- the risks to the health and safety of persons not in their employment arising out of or in connection with their work activities.

It is vital that you know how to **carry out a risk assessment**. Often you may be in a position where you are given direct responsibility for this, and the care and attention you take over it may have a direct impact on the safety of others. You must be aware of the dangers or hazards of any task, and know what can be done to prevent or reduce the risk.

There are five steps in a risk assessment:

- **Step 1** Identify the hazards
- **Step 2** Identify who is at risk
- **Step 3** Calculate the risk from the hazard against the likelihood of it taking place
- **Step 4** Introduce measures to reduce the risk
- **Step 5** Monitor the risk.

Method statements

Method statements are key safety documents that take the information about significant risks from your risk assessment, and combine them with the job specification, to produce a practical and safe working method for the workers to follow for tasks on site.

Hazard books

The hazard book is a tool used on some sites that identifies hazards within certain tasks and can help to produce risk assessments. The book will list tasks and what hazards are associated with those tasks.

K4. Health and hygiene

As well as keeping an eye out for hazards, you must also make sure that you look after yourself and stay healthy. This is a responsibility that lies with both the employer and the employee.

Staying healthy

One of the easiest ways to stay healthy is to wash your hands on a regular basis. By washing your hands you are preventing hazardous substances from entering your body through ingestion (swallowing). You should always wash your hands after going to the toilet and before eating or drinking. Personal hygiene is vital to ensure good health.

Remember that some health problems do not show symptoms straight away and what you do now can affect you much later in life.

Welfare facilities

Welfare facilities are things such as toilets, which must be provided by your employer to ensure a safe and healthy workplace. There are several things that your employer must provide to meet welfare standards:

- toilets
- washing facilities
- drinking water
- storage room for clothes and personal belongings
- lunch areas.

Substance abuse

Substance abuse is a general term and mainly covers things such as drinking alcohol and taking drugs. Taking drugs or inhaling solvents at work is not only illegal, but is also highly dangerous to

Remember

There may be different hazards that are associated with tasks. Different working environments can create different types of hazard so risk assessments must always look at the specific task and not a generic one

Figure 1.5 Always wash your hands to prevent ingesting hazardous substances

Safety tip

When placing clothes in a drying room, do not place them directly onto heaters as this can lead to fire

Remember

Prescription drugs can also affect you on site – be sure to follow all instructions clearly

Did you know?

Noise is measured in decibels (dB). The average person may notice a rise of 3dB, but with every 3dB rise, the noise is doubled. What may seem like a small rise is actually very significant

Remember

Reducing the risk of noise damage can be done in a number of ways:

Remove – get rid of whatever is creating the noise

Move – locate noisy equipment away from people

Enclose – surround noisy equipment, e.g. with sound proof material

Isolation – move the workers to protected areas

Even after all of these are considered, PPE may still be required

Safety tip

Not all substances are labelled, and sometimes the label may not match the contents. If you are in any doubt, don't use or touch the substance

you and everyone around you as reduced concentration problems can lead to accidents. Drinking alcohol is also dangerous at work; going to the pub for lunch and having just one drink can lead to slower reflexes and reduced concentration.

Health effects of noise

Damage to hearing has a range of causes, from ear infections to loud noises. Hearing loss can result from one very loud noise lasting only a few seconds, or from relatively loud noise lasting for hours, such as a drill.

The damage to hearing can be caused by one of two things:

- **intensity** – you can be hurt in an instant from an explosive or very loud noise which can burst your eardrum
- **duration** – noise doesn't have to be deafening to harm you, it can be a quieter noise over a long period, e.g. a 12 hour shift.

Hazardous substances

Hazardous substances are a major health and safety risk on a construction site. To this end, they need to be handled, stored, transported and disposed of in very specific ways.

- Step 1 – assess the risks to health from hazardous substances used or created by employees' activities
- Step 2 – decide what precautions are needed
- Step 3 – prevent employees from being exposed to any hazardous substances. If prevention is impossible, the risk must be adequately controlled
- Step 4 – ensure control methods are used and maintained properly.
- Step 5 – monitor the exposure of employees to hazardous substances
- Step 6 – carry out health surveillance to ascertain if any health problems are occurring
- Step 7 – prepare plans and procedures to deal with accidents such as spillages
- Step 8 – ensure all employees are properly informed, trained and supervised.

Identifying a substance that may fall under the COSHH regulations is not always easy, but you can ask the supplier or manufacturer for a COSHH data sheet, outlining the risks involved with a substance. Most substance containers carry a warning sign stating whether the contents are corrosive, harmful, toxic or bad for the environment.

Figure 1.6 Common safety signs for corrosive, toxic and explosive materials

Waste

Many different types of waste material are produced in construction work. It is your responsibility to identify the type of waste you have created and the best way of disposing it.

There are several pieces of legislation that dictate the disposal of waste materials. They include:

- Environmental Production Act 1990
- Controlled Waste Regulations 1992
- Waste Management Licensing Regulations 1994.

Several different types of waste are defined by these regulations:

- household waste – normal household rubbish
- commercial waste – from shops or offices
- industrial waste – from factories and industrial sites.

All waste must be handled properly and disposed of safely. The Controlled Waste Regulations state that only those authorised to do so may dispose of waste and that a record should be kept of all waste disposal.

Hazardous waste

Some types of waste such as chemicals or material that is toxic or explosive, are too dangerous for normal disposal and must be disposed of with special care. The Hazardous Waste Regulations (England and Wales) cover this disposal. If hazardous material is inside a container the container must be clearly marked and a consignment note completed for its disposal.

Health risks in the workplace

While working in the construction industry, you will be exposed to substances or situations that may be harmful to your health. Some of these health risks may not be noticeable straight away. It may take years for **symptoms** to be noticed and recognised.

Remember

If you leave material on site when your work is completed you may be discarding them. You are still responsible for this waste material!

Key terms

Symptom – a sign of illness or disease (e.g. difficulty breathing, a sore hand or a lump under the skin)

Leptospirosis – an infectious disease that affects humans and animals. The human form is commonly called Weil's disease. The disease can cause fever, muscle pain and jaundice. In severe cases it can affect the liver and kidneys. Leptospirosis is a germ that is spread by the urine of the infected person. It can often be caught from contaminated soil or water that has been urinated in

Dermatitis – a skin condition where the affected area is red, itchy and sore

Vibration white finger – a condition that can be caused by using vibrating machinery (usually for very long periods of time). The blood supply to the fingers is reduced which causes pain, tingling and sometimes spasms (shaking)

Remember

Activities on site can damage your body. You could have eye damage, head injury and burns along with other physical wounds

Ill-health can result from:

- exposure to dust (such as asbestos), which can cause eye injuries, breathing problems and cancer
- exposure to solvents or chemicals, which can cause **leptospirosis**, **dermatitis** and other skin problems
- lifting heavy or difficult loads, which can cause back injury and pulled muscles
- exposure to loud noise, which can cause hearing problems and deafness
- exposure to sunlight, which can cause skin cancer
- using vibrating tools, which can cause **vibration white finger** and other problems with the hands
- head injuries, which can lead to blackouts and epilepsy
- cuts, which if infected can lead to disease.

Everyone has a responsibility for health and safety in the construction industry but accidents and health problems still happen too often. Make sure you do what you can to prevent them.

K5. Safe handling of materials and equipment

Manual handling

Manual handling means lifting and moving a piece of equipment or material from one place to another without using machinery. Lifting and moving loads by hand is one of the most common causes of injury at work.

Poor manual handling can cause injuries such as muscle strain, pulled ligaments and hernias. The most common injury by far is spinal injury. Spinal injuries are very serious because there is very little doctors can do to correct them and, in extreme cases, workers have been left paralysed.

Safety tip

Most injuries caused by manual handling result from years of lifting items that are too heavy, are awkward shapes or sizes, or from using the wrong technique. However, it is also possible to cause a lifetime of back pain with just one single lift

Lifting correctly (kinetic lifting)

When lifting any load it is important to keep the correct posture and to use the correct technique.

The correct posture before lifting:

- feet should be shoulder-width apart with one foot slightly in front of the other
- knees should be bent
- back must be straight
- arms should be as close to the body as possible
- grip must be firm using the whole hand and not just the fingertips.

The correct technique when lifting:

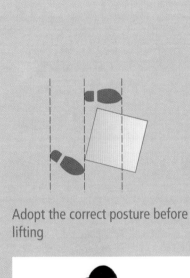

Adopt the correct posture before lifting

Get a good grip on the load

Adopt the correct posture when lifting

Move smoothly with the load

Adopt the correct posture and technique when lowering

Unit 1001 Safe working practices in construction

- approach the load squarely facing the direction of travel
- adopt the correct posture
- place hands under the load and pull the load close to your body
- lift the load using your legs and not your back.

When lowering a load you must also adopt the correct posture and technique:

- bend at the knees, not the back
- adjust the load to avoid trapping fingers
- release the load.

Safe handling

Safe manual handling methods are discussed in detail on page 19. When handling any materials or equipment, always think about the health and safety issues involved and remember the manual handling practices explained to you during your induction.

You aren't expected to remember everything but basic common sense will help you to work safely.

- Always wear your safety helmet and boots at work.
- Wear gloves and ear defenders when necessary.
- Keep your work areas free from debris, materials, tools and equipment not being used.
- Wash your hands before eating.
- Use barrier cream before starting work.
- Always use correct lifting techniques.

Did you know?

In 2004/2005 there were over 50 000 injuries while handling, lifting or carrying in the UK (Source: HSE)

Ensure you follow instructions given to you at all times when moving any materials or equipment. The main points to remember are:

- Always try to avoid manual handling (or use mechanical means to aid the process).
- Always assess the situation first to decide on the best method of handling the load.
- Always reduce risks as much as possible (e.g. split a very heavy load and move obstacles from your path before lifting).
- Tell others around you what you are doing.
- If you need help with a load, get it. Don't try to lift something beyond what you can manage.

Always treat power tools with respect: they have the potential to cause harm either to the person using them or to others around. All power tools used on site should be regularly tested (**PAT** tested) by a qualified person. There are several health and safety regulations governing the use of power tools. Make sure that you wear suitable PPE at all times and that power tools are operated safely. In some cases, you must be qualified to use them. Refer to PUWER (Provision and Use of Work Equipment Regulations 1998) if needed.

On-site transformers are used to reduce the mains voltage from 230 volts to 110 volts. All power tools used should be designed for 110 volts.

As well as traditional power tools there are also tools powered by gas or compressed air. Gas-powered tools such as nail guns, also require batteries to operate them. They must be handled carefully similar to other power tools.

Special care should be taken with electrical tools.

ALWAYS:

- check plugs and connections (make sure you have the correct fuse rating in the plug)
- inspect all leads to ensure there is no damage
- check that the power is off when connecting leads
- unwind extension leads completely from the reel to prevent the cable from overheating.

NEVER:

- use a tool in a way not recommended by the manufacturer
- use a tool with loose, damaged or makeshift parts
- lay a driver down while it is still switched on
- use a drill unless the chuck (the part in which the drill bit is held) is tight
- throw the tool onto the ground
- pass the tool down by its lead
- use a drill where it is difficult to see what you are doing or to hold the tool tightly
- allow leads to trail in water.

Unit 1001 Safe working practices in construction

Safety tips

- Hand tools with sharp edges should be covered to prevent cuts
- Power tools should be carried by the handle
- Power tools that have gas-powered cartridges must be stored in an area that is safe and away from sources of ignition to prevent explosion. Used cartridges must be disposed of safely

Remember

Guidance for the storage of power tools will be provided in the manufacturer's manual. Most power tools come with a plastic carry case that can be used for storage

Functional skills

To store materials safely and correctly, you will need to be familiar with manufacturers' instructions. In doing this you will be practising the following FE 1.2.3: Reading different texts and taking appropriate action.

Safety tip

It is good practice to put an intermediate flat stack in long rows to prevent rows from toppling

Safe storage and handling of tools and equipment

Tools

All tools need to be stored safely and securely in suitable bags or boxes to protect them from weather and rust. When not in use they should be safely locked away.

Bricks and blocks

Type	Storage and handling issues
Bricks	• Now largely pre-packed and banded using either plastic or metal bands to stop bricks separating until ready for use. Edges are protected by plastic strips to prevent damage during moving. Usually covered in shrink-wrapped plastic to protect from the elements. • Store on level ground close to where they are required and stack no more than two packs high. • Once banding is cut, bricks can collapse, so great care must be taken. • Bricks should be taken from a minimum of three packs and mixed to avoid changes in colour. If bricks are not mixed different shades will appear in brickwork – this is called banding. • If unloaded by hand the bricks should be stacked on edge in rows with the ends of stacks bonded and no higher than 1.8 m. All stacks should be covered with tarpaulin or polythene sheets.
Blocks	Made from concrete and may be dense or lightweight. Storage is the same as bricks.
Paving slabs	• Made from concrete or stone in a variety of shapes and sizes. • Normally delivered by lorry in wooden crates, covered in shrink-wrapped plastic, or banded and covered on pallets. Do not stack higher than two packs to help prevent accidents and weight pressure on slabs which can cause damage. • Store outside and stack on edge to prevent lower slabs being damaged by the weight of the stack. Smaller numbers can be stored flat. Store on firm, level ground with timber bearers below to prevent damage to edges.

Aggregates

Aggregates are granules or particles that are mixed with cement and water to make mortar and concrete. Aggregates should be hard, durable and should not contain any form of plant life, or anything that can be dissolved in water.

Aggregates are normally delivered in tipper lorries, although nowadays one-tonne bags are available and may be crane handled. The aggregates should be stored on a concrete base, with a fall to allow for any water to drain away. In order to protect aggregates from becoming contaminated with leaves and litter, it is a good idea to situate stores away from trees and cover aggregates with a tarpaulin or plastic sheets.

Cement and plaster

Plaster is made from gypsum, water and cement or lime. Aggregates can also be added depending on the finish desired. Plaster provides a jointless, smooth, easily decorated surface for internal walls and ceilings. Cement is made from limestone or chalk and is used in the creation of mortar and concrete.

Both cement and plaster are usually available in 25 kg bags. The bags are made from multi-wall layers of paper with a polythene liner. Care must be taken not to puncture the bags before use. Each bag, if offloaded manually, should be stored in a ventilated, waterproof shed, on a dry floor on pallets. If offloaded by crane they should be transferred to the shed and the same storage method should be used.

The bags should be kept clear of the walls, and piled no higher than five bags.

Wood and sheet materials

There are various types of wood and sheet materials available, but the most common are as follows:

Type	Storage and handling issues
Carcassing timber	Store outside under a covered framework on timber bearers clear of ground, which should be vegetation-free, to reduce the risk of ground moisture absorption. Use piling sticks between each layer of timber to provide support and allow air circulation.
Joinery grade and hardwoods	Store under full cover, preferably in a storage shed. Good ventilation is needed to prevent a build-up of moisture. Store on bearers on a well-prepared base.

Remember

Aggregates can be supplied as bagged materials, as can cement and plaster

Did you know?

On larger sites some companies use a machine spray system to cover large areas with plaster quickly, using many plasterers to complete the work

Type	Storage and handling issues
Plywood and sheet materials	Store in a dry, well-ventilated environment. Specialised covers are available to give added protection, helping to prevent condensation from other types of sheeting. Stack flat on timber cross-bearers, spaced close together to prevent sagging. Where space is short, store on edge in purpose-made racks. There should be sufficient space for easy loading and removal. Do not lean against walls as this makes the wood bow.
	Sheet materials with faces should be placed against each other to minimise the risk of damage. Keep different sizes, grades and qualities of sheet materials separate.
	Sheet materials are heavy and easy to damage so special care is needed when moving them.
Joinery components	These can be doors, kitchen units, etc. and must be stored safely and securely.
	Doors, frames, etc. should ideally be stored flat on timber bearers under cover to protect from the weather. In limited space they can be stored upright in a rack, but do not lean against a wall.
	Wall and floor units must be stacked on a flat surface no more than two units high. Store inside, preferably in the room they are to be fitted in. Use protective sheeting to prevent damage and staining.
Plasterboard	Larger sheets can be very awkward to carry, particularly in strong wind. Store in a flat waterproof area and do not lean against a wall.

Adhesives

Adhesives are substances used to bond (stick) surfaces together. Because of their chemical nature, there are a number of potentially serious risks connected with adhesives if they are not stored, used and handled correctly.

Figure 1.7 Adhesives should be stored according to the manufacturer's instructions

Remember

Heavy materials should be stored at low levels to aid manual handling and should never be stacked more than two levels high

Figure 1.8 Correct storage of paints

All adhesives should be stored and used in line with the manufacturer's instructions. This usually involves storing them on shelving, with labels facing outwards, in a safe, secure area (preferably a lockable storeroom). It is important to keep the labels facing outwards so that the correct adhesive can be selected.

Paint and decorating equipment

Type	Storage issues
Oil-based	Store on clearly marked shelves with the labels turned to the front. Always order in date order, with new stock at the back. They should be regularly **inverted** to prevent settlement or separation of ingredients and kept tightly sealed to prevent **skinning**. Store at a constant temperature to maintain consistency.
Water based	Store on clearly marked shelves with the labels turned to the front and in date order. Store at a constant temperature and protect from frost to prevent freezing. Use before the use-by-date expires.
Powdered materials	Heavy bags should be stored at ground or platform level. Smaller items can be stored on shelves with loose materials in sealed containers. Protect from frost and moisture and from high humidity. Do not store in the open air.

Key terms

Inverted – tipped and turned upside down

Skinning – the formation of a skin which occurs when the top layer dries out

Substances hazardous to health

Some substances a decorator will work with are potentially hazardous to health, with **volatile** and highly flammable characteristics. The Control of Substances Hazardous to Health (COSHH) Regulations apply to such materials and detail how they must be stored and handled.

Decorating materials that might be hazardous to health include spirits (i.e. methylated and white), turpentine (turps), paint thinners and varnish removers. These should be stored out of the way on shelves, preferably in a suitable locker or similar room that meets the requirements of COSHH. The temperatures must be kept below 15°C as a warmer environment may cause storage containers to expand and blow up.

Key term

Volatile – quick to evaporate (turn into a gas)

Figure 1.9 A harness and lanyard can prevent a worker from falling to the ground

K6. Basic working platforms

Fall protection

With any task that involves working at height, the main danger to workers is falling. Although scaffolding, etc. should have edge protection to prevent falls, there are certain tasks where edge protection or scaffolding simply cannot be used.

In these instances some form of fall protection must be in place to prevent the worker falling, keep the fall distance to a minimum or ensure the landing point is cushioned.

Harnesses and lanyards

Harnesses and lanyards are a type of fall-arrest system, which means that, in the event of a slip or fall, the worker will only fall a few feet at most.

The system works with a harness that is attached to the worker and a lanyard attached to a secure beam/eyebolt. If the worker slips, they will only fall as far as the length of cord/lanyard and will be left hanging, rather than falling to the ground.

Safety netting

Safety netting is also a type of fall-arrest system but is used mainly on the top floor where there is no higher point to attach a lanyard. The nets are attached to the joists/beams and are used to catch any worker who may slip or fall.

Figure 1.10 Safety netting is used when working at the highest point

Airbags

An airbag safety system is a form of soft fall-arrest and is comprised of interlinked modular air mattresses that expand together to form a continuous protective safety surface, giving a cushioned fall and preventing serious injury.

The system must be kept inflated and should be checked regularly to ensure that it is still functional. This system is ideal for short-fall jobs, but should not be used where a large fall could occur.

Stepladders and ladders

Stepladders

A stepladder has a prop, which when folded out allows the ladder to be used without having to lean it against something. Stepladders are one of the most frequently used pieces of access equipment in the construction industry and are often used every day. This means that they are not always treated with the respect they demand.

Stepladders are often misused. They should only be used for work that will take a few minutes to complete. When work is likely to take longer, use a sturdier alternative.

When stepladders are used, the following safety points should be considered:

- Ensure the ground on which the stepladder is to be placed is firm and level. If the ladder rocks or sinks into the ground it should not be used.
- Always open the steps fully.
- Never work off the top tread of the stepladder.
- Always keep your knees below the top tread.
- Never use stepladders to gain additional height on another working platform.
- Always look for the kitemark (Figure 1.11), which shows that the ladder has been made to BSI standards.

A number of other safety points need to be considered depending on the type of stepladder being used.

Wooden stepladder

Before using a wooden stepladder, you should check:

- for loose screws, nuts, bolts and hinges
- that the tie ropes between the two sets of **stiles** are in good condition and not frayed

Figure 1.11 British Standards Institution kitemark

Key term

Stiles – the side pieces of a stepladder into which the steps are set

Safety tip

If any faults are revealed when checking a stepladder, it should be taken out of use, reported to the person in charge and a warning notice attached to it to stop anyone using it

Find out

What are the advantages and disadvantages of each type of stepladder?

Did you know?

Ladders and stepladders should be stored under cover to protect them from damage such as rust or rotting

- for splits or cracks in the stiles
- that the treads are not loose or split.

Never paint any part of a wooden stepladder as this can hide defects, which may cause the ladder to fail during use, causing injury.

Aluminium stepladder

Before using an aluminium stepladder:

- check for damage to stiles and treads to see whether they are twisted, badly dented or loose
- ensure you are not working close to live electricity supplies as aluminium will conduct electricity.

Fibreglass stepladder

Before using a fibreglass stepladder, check for damage to stiles and treads. Once damaged, fibreglass stepladders cannot be repaired and must be disposed of.

Ladders

A ladder, unlike a stepladder, does not have a prop and has to be leant against something in order for it to be used. Together with stepladders, ladders are one of the most common pieces of equipment used to carry out work at height and to gain access to the work area.

Ladders are also made of timber, aluminium or fibreglass, and require similar checks to stepladders before use.

Erecting and using a ladder

The following points should be noted when considering the use of a ladder:

- As with stepladders, ladders are not designed for work of long duration. Alternative working platforms (see pages 29–32) should be considered if the work will take longer than a few minutes.
- The work should not require the use of both hands. One hand should be free to hold the ladder.
- You should be able to do the work without stretching.
- You should make sure that the ladder can be adequately secured to prevent it slipping on the surface it is leaning against.

Pre-use checks

Before using a ladder check its general condition. Make sure that:

- no rungs are damaged or missing
- the stiles are not damaged

- no **tie-rods** are missing
- no repairs have been made to the ladder.

In addition, for wooden ladders ensure that:

- they have not been painted, which may hide defects or damage
- there is no decay or rot
- the ladder is not twisted or warped.

Erecting a ladder

Consider the following guidelines when erecting a ladder:

- Ensure you have a solid, level base.
- Do not pack anything under either (or both) of the stiles to level it.
- If the ladder is too heavy to put it in position on your own, get someone to help.
- Ensure that there is at least a four-rung overlap on each extension section.
- Never rest the ladder on plastic guttering as it may break, causing the ladder to slip and the user to fall.
- Where the base of the ladder is in an exposed position, ensure it is adequately guarded so that no one knocks it or walks into it.
- The ladder should be secured at both the top and bottom. The bottom of the ladder can be secured by a second person, however, this person must not leave the base of the ladder while it is in use.
- The angle of the ladder should be a ratio of 1:4 (or 75°). This means that the bottom of the ladder is 1 m away from the wall for every 4 m in height (see Figure 1.12).
- The top of the ladder must extend at least 1 m, or five rungs, above its landing point.

Trestle platforms

Frames

A-frames

These are most commonly used by carpenters and painters. As the name suggests, the frame is in the shape of a capital A and can be made from timber, aluminium or fibreglass. Two are used together to support a platform (a scaffold or staging board).

When using A-frames:

- they should always be opened fully and, in the same way as stepladders, must be placed on firm, level ground

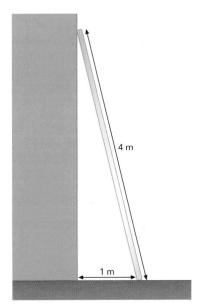

Figure 1.12 Correct angle for a ladder

- the platform width should be no less than 450 mm
- the overhang of the board at each end of the platform should not be more than four times its thickness.

Steel trestles

These are sturdier than A-frame trestles and are adjustable in height. They are also capable of providing a wider platform than timber trestles. As with the A-frame type, they must be used only on firm and level ground but the trestle itself should be placed on a flat scaffold board on top of the ground. Trestles should not be placed more than 1.2 m apart.

Platforms

Scaffold boards

To ensure that scaffold boards provide a safe working platform, check that they:

- are not split
- are not twisted or warped
- have no large knots, which cause weakness.

Care should be taken when handling scaffold boards as they can be long and unwieldy. Ideally two people should carry them. It is important to store scaffold boards correctly, that is flat and level, otherwise they will twist or bow. They also need to be kept covered to prevent damage from rain, which could lead to rot.

Staging boards

These are designed to span a greater distance than scaffold boards and can offer a 600 mm wide working platform. They are ideal for use with trestles.

Hop-ups

Also known as step-ups, hop-ups are ideal for reaching low-level work that can be carried out in a relatively short period of time. A hop-up needs to be of sturdy construction and have a base of not less than 600 mm by 500 mm. Hop-ups have the disadvantage that they are heavy and awkward to move around.

Scaffolding

Tubular scaffold is the most commonly used type of scaffolding in the construction industry. There are two types of tubular scaffold:

- **independent scaffold** – free-standing scaffold that does not rely on any part of the building to support it (although it must be tied to the building to provide additional stability)

> **Safety tip**
>
> A-frame trestles should never be used as stepladders as they are not designed for this purpose

> **Safety tip**
>
> Do not use items as hop-ups that are not designed for the purpose (e.g. milk crates, stools or chairs). They are usually not very sturdy and can't take the weight of someone standing on them. This may result in falls and injury

- **putlog scaffold** – scaffolding that is attached to the building via the entry of some of the poles into holes left in the brickwork by the bricklayer. The poles stay in position until the construction is complete and give the scaffold extra support.

No one other than a qualified **carded scaffolder** is allowed to erect or alter scaffolding. Although you are not allowed to erect or alter this type of scaffold, you must be sure it is safe before you work on it.

Mobile tower scaffolds

Mobile tower scaffolds are so called because they can be moved around without being dismantled. Lockable wheels make this possible and they are used extensively throughout the construction industry by many different trades. A tower can be made from either traditional steel tubes and fittings or aluminium, which is lightweight and easy to move. The aluminium type tower is normally specially designed and is referred to as a 'proprietary tower'.

A 'low tower' is a different type of tower scaffold. These are designed to be used by one person and have a recommended working height of 2.5 m and a working load of 150 kg. They need no assembly beyond locking in place the platform and handrails, but make sure you follow the manufacturer's instructions to do this.

Erecting a tower scaffold

It is essential that tower scaffolds are situated on a firm and level base. The stability of any tower depends on the height in relation to the size of the base:

- for use inside a building, the height should be no more than three and a half times the smallest base length
- for outside use, the height should be no more than three times the smallest base length.

The height of a tower can be increased provided the area of the base is increased **proportionately**. The base area can be increased by fitting outriggers to each corner of the tower.

For mobile towers, the wheels must be in the locked position whilst they are in use and unlocked only when they are being repositioned.

There are several important points you should consider when working from a scaffold tower:

- any working platform above 2 m high must be fitted with guardrails and toe boards. Guardrails may also be required at heights of less than 2 m if there is a risk of falling onto potential

Key term

Carded scaffolder – someone who holds a recognised certificate showing competence in scaffold erection

Remember

If you have any doubts about the safety of scaffolding, report them. You could very well prevent serious injury or even someone's death

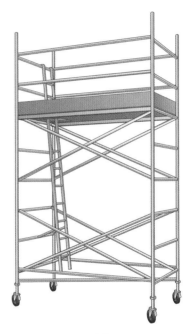

Figure 1.13 Mobile tower scaffold

Key term

Proportionately – in relation to the size of something else

Safety tip

Mobile towers must only be moved when they are free of people, tools and materials

Safety tip

Never climb a scaffold tower on the outside as this can cause it to tip over

hazards below, for example reinforcing rods. Guardrails must be fitted at a minimum height of 950 mm

- if guardrails and toe boards are needed, they must be positioned on all four sides of the platform
- any tower higher than 9 m must be secured to the structure
- towers must not exceed 12 m in height unless they have been specifically designed for that purpose
- the working platform of any tower must be fully boarded and be at least 600 mm wide
- if the working platform is to be used for materials, the minimum width must be 800 mm
- all towers must have their own access and this should be by an internal ladder.

The dangers of working at height

As well as falling from the height, there are additional dangers in working at height that are not present when working at ground level.

Although good housekeeping is important while working at ground level to prevent slips and trips, it is *vital* when working at height. Not only are you at added risk, but materials and tools that are left on a working platform can be knocked off the platform onto people working below. There is a risk of causing serious head injuries to those below – and not just the workforce, as in some instances the working platform may be in an area that involves the general public.

When working in a public area the public must be protected from hazards by way of barriers around the work area. You must also ensure that the sides of the working platform are sealed off to prevent any materials or other objects from falling.

Working life

Ralph and Vijay are working on the second level of some scaffolding, clearing some debris. Ralph suggests that, to speed up the task, they should throw the debris over the edge of the scaffolding into the skip below. The building Ralph and Vijay are working on is on a main road and the skip is not in a closed off area.

What do you think about Ralph's idea? Ralph seems to have forgotten that working at height poses a risk not only to himself and Vijay, but also to the people below them. What might be the effects of following Ralph's plan? The scaffold should be set up to prevent materials from going over the edge, so to follow the plan will already mean having to make the scaffold less safe for Ralph and Vijay.

What should Ralph and Vijay be doing to clear the debris from the scaffold?

K7. Working with electricity

Electricity is a killer. One of the main problems with electricity is that it is invisible. You don't even have to be working with an electric tool to be electrocuted. You can get an electric shock:

- working too close to live overhead cables
- plastering a wall with electric sockets
- carrying out maintenance work on a floor
- drilling into a wall.

Voltages

There are two main types of voltage in use in the UK. These are 230 V and 110 V. The standard UK power supply is 230 V and this is what all the sockets in your house are.

230 V has been deemed unsafe on construction sites so 110 V must be used here. The 110 V is identified by a yellow cable and different style plug. It works from a transformer which converts the 230 V to 110 V.

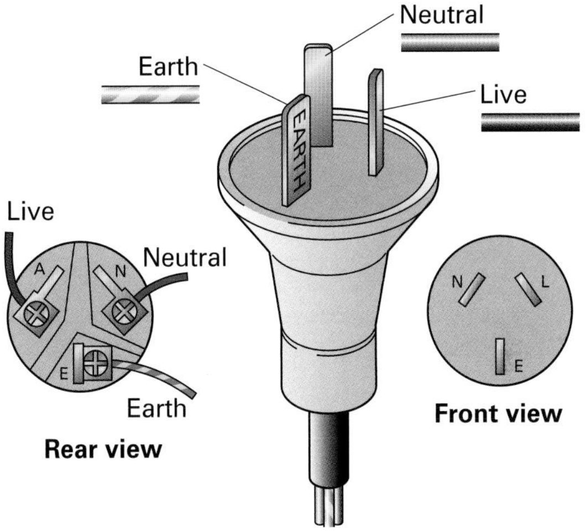

When working within domestic dwellings where 230 V is the standard power source, a portable transformer should ideally be used. If this is not possible, residual current devices (RCD) should be used.

Contained within the wiring there should be three wires: the live and neutral, which carry the alternating current, and the earth wire, which acts as a safety device. The three wires are colour-coded as follows to make them easy to recognise:

Live – Brown
Neutral – Blue
Earth – Yellow and green

Precautions to take to prevent electric shocks

NEVER:

- carry electrical equipment by the cable
- remove plugs by pulling on the lead
- allow tools to get wet. If they do, get them checked before use.

Did you know?

Around 30 workers a year die from electricity-related accidents, with over 1000 more being seriously injured (source: HSE)

Figure 1.14 Colour coding of the wires in a 230 V-plug

Figure 1.15 A 110 V-plug

Safety tip

The colour coding of the wires has been changed recently to comply with European colours. Some older properties will have the following colours:
Live – Red
Neutral – Black
Earth – Yellow and green

Unit 1001 Safe working practices in construction

ALWAYS:

- check equipment, leads and plugs before use. If you find a fault don't use the equipment and tell your supervisor immediately
- keep cables off the ground where possible to avoid damage/ trips
- avoid damage to the cable by keeping it away from sharp edges
- keep the equipment locked away and labelled to prevent it being used by accident
- use cordless tools where possible
- follow instructions on extension leads.

Dealing with electric shocks

In helping a victim of an electric shock, the first thing you must do is disconnect the power supply – if it is safe to do and will not take long to find. Touching the power source may put you in danger.

- If the victim is in contact with something portable such as a drill, attempt to move it away using a non-conductive object such as a wooden broom.
- Time is precious and separating the victim from the source can prove an effective way to speed the process.
- Don't attempt to touch the affected person until they are free and clear of the supplied power. Be especially careful in wet areas such as bathrooms – water conducts electricity and electrocuting yourself is a possibility.

People 'hung up' in a live current flow may think they are calling out for help but most likely no sound will be heard from them. When the muscles contract under household current (most electrocutions happen from house current at home), the person affected will appear in 'locked-up' state, unable to move or react to you.

- Using a wooden object, swiftly and strongly knock the person free, trying not to injure them, and land them clear of the source.
- The source may also be lifted or removed, if possible, with the same wooden item. This is not recommended on voltages that exceed 500 V.

K8. Using appropriate PPE

Personal protective equipment is the name for clothes and other wearable items that form a line of defence against accidents or injury. PPE is not the only way of preventing accidents or injury. It should be used with all the other methods of staying healthy and safe in the workplace (equipment, training, regulations and laws, etc.).

Maintaining and storing PPE

It is important that PPE is well-maintained. The effectiveness of the protection it offers will be affected if the PPE is damaged in any way. Maintenance may include:

- cleaning
- examination
- replacement
- repair and testing.

The wearer may be able to carry out simple maintenance (such as cleaning), but more intricate repairs must only be carried out by a competent person. The costs associated with the maintenance of PPE are the responsibility of the employer.

Where PPE is provided, adequate storage facilities for PPE must also be provided for when it is not in use, unless the employee may take PPE away from the workplace (e.g. footwear or clothing).

Accommodation may be simple (e.g. pegs for waterproof clothing or safety helmets) and it need not be fixed (e.g. a case for safety glasses or a container in a vehicle). Storage should be adequate to protect the PPE from contamination, loss, damage, damp or sunlight. Where PPE may become contaminated during use, storage should be separate from any storage provided for ordinary clothing.

PPE should be 'CE' marked. This will indicate that it complies with the requirements of the Personal Protective Equipment Regulations 2002 (see page 8). The CE marking shows that the PPE meets safety requirements. In some cases it may have been tested and certified by an independent body.

The possible consequences of not using PPE can be serious and cause long-term health problems. The health problems and their consequences are described on page 18.

Remember

PPE only works properly if it is being used and used correctly! The main pieces of legislation that govern the use of PPE are:
- Control of Substances Hazardous to Health 2002
- Provision and Use of Work Equipment Regulations (1992 and 1998)
- Personal Protective Equipment at Work Regulations 1992

Unit 1001

Safe working practices in construction

Types of PPE

Head protection

The most common piece of head protection used in construction is the safety helmet (or hard hat). This protects the head from falling objects and knocks and has an adjustable strap to ensure a snug fit.

Figure 1.16 A safety helmet

Eye protection

Eye protection is used to protect the eyes from dust and flying debris. The three main types are:

- **safety goggles** – made of a durable plastic and used when there is a danger of dust getting into the eyes or a chance of impact injury
- **safety spectacles** – these are also made from a durable plastic but give less protection than goggles. This is because they don't fully enclose the eyes and so only protect from flying debris
- **facemasks** – again made of durable plastic, facemasks protect the entire face from flying debris. They do not, however, protect the eyes from dust.

Figure 1.17 Safety goggles

Figure 1.18 Safety spectacles

Foot protection

Safety boots or shoes are used to protect the feet from falling objects and to prevent sharp objects such as nails from injuring the feet. Safety boots should have a steel toe-cap and steel mid-sole.

Hearing protection

Hearing protection is used to prevent damage to the ears caused by very loud noise. The two most common types are earplugs and ear defenders.

- **ear-plugs** – these are small fibre plugs that are inserted into the ear and used when the noise is not too severe
- **ear defenders** – these are worn to cover the entire ear to protect from excessive noise and are connected to a band that fits over the top of the head.

Figure 1.19 Safety boots

Figure 1.20 Earplugs

Figure 1.21 Ear defenders

Figure 1.22 A respiratory system

Figure 1.23 Safety gloves

Respiratory protection

Respiratory protection is used to prevent the worker from breathing in any dust or fumes that may be hazardous. The main type of respiratory protection is the dust mask.

Dust masks are used when working in a dusty environment and are lightweight, comfortable and easy to fit.

Hand protection

There are several types of hand protection and the correct type should be used for each task. For example, wearing lightweight rubber gloves to move glass will not offer much protection, so leather gauntlets must be used. Plastic-coated gloves will protect you from certain chemicals and Kevlar® gloves offer cut resistance.

Skin and sun protection

Another precaution you can take is ensuring that you wear barrier cream. This is a cream used to protect the skin from damage and infection. Don't forget to ensure that your skin is protected from the sun with a good sunscreen, and make sure your back, arms and legs are covered by suitable clothing.

Whole body protection

The rest of the body also needs protecting when working on site. This will usually involve either overalls which protect from dirt and minor cuts, or high-visibility jackets which make the wearer visible at all times.

Safety tip

Dust masks only offer protection from non-toxic dust, so if the worker is to be exposed to toxic dust or fumes, a full respiratory system should be used

K9. Fire and emergency procedures

Fires can start almost anywhere and at any time, but a fire needs all the ingredients of 'the triangle of fire' to burn. Remove one side of the triangle, and the fire will be extinguished. Fire moves by consuming all these ingredients and burns fuel as it moves.

Figure 1.24 The triangle of fire

Fires can be classified according to the type of material that is involved:

- Class A – wood, paper, textiles, etc.
- Class B – flammable liquids, petrol, oil, etc.
- Class C – flammable gases, liquefied petroleum gas (LPG), propane, etc.
- Class D – metal, metal powder, etc.
- Class E – electrical equipment.

Fire-fighting equipment

There are several types of fire-fighting equipment such as fire blankets and fire extinguishers.

Fire extinguishers

A fire extinguisher is a metal canister containing a substance that can put out a fire.

Fire blankets

Fire blankets are made of a fireproof material and smother the fire stopping any more oxygen from getting to it. A fire blanket can be used if a person is on fire.

What to do in the event of a fire

During your induction to any workplace, you will be made aware of the fire procedure as well as where the fire assembly points (also known as muster points) are and what the alarm sounds like. On hearing the alarm you must stop what you are doing and make your way calmly to the nearest muster point.

Find out

What fire risks are there in the construction industry? Think about some of the materials (fuel) and heat sources that could make up two sides of 'the triangle of fire'

Remember

- Remove the fuel – without anything to burn, the fire will go out
- Remove the heat and the fire will go out
- Remove the oxygen and the fire will go out – without oxygen, a fire won't even ignite

Safety tip

It is important to remember that when you put out a fire with a fire blanket, you must take extra care as you will have to get quite close to the fire

Figure 1.25 A fire blanket

Fire extinguisher	Colour Band	Main use	Details
Water fire extinguisher	Red	Class A fires	Never use it for an electrical or burning fat/oil fire, as water can conduct the electricity back to the person using the extinguisher. Putting water on oil or fat fires will 'explode' the fire, making it worse.
Foam fire extinguisher	Cream	Class A fires	This can also be used on Class B if no liquid is flowing and on Class C if gas is in liquid form.
Carbon dioxide (CO_2) extinguisher	Black	Class E	Primarily used on electrical fires; can also be used in Class A, B and C.
Dry powder extinguisher	Blue	All classes	Most commonly used on electrical and liquid fires. The powder puts out the fire by smothering the flames.

K10. Safety signs and notices

There are many different safety signs but each will usually fit into one of four categories:

1. **Prohibition signs** – these tell you that something MUST NOT be done. They always have a white background and a red circle with a red line through it.

Figure 1.26 A prohibition sign

2. **Mandatory signs** – these tell you that something MUST be done. They are also circular, but have a white symbol on a blue background.

Figure 1.27 A mandatory sign

3. **Warning signs** – these signs are there to alert you to a specific hazard. They are triangular and have a yellow background and a black border.

Figure 1.28 A warning sign

4. **Safe condition signs** (often called information signs) – these give you useful information. They can be square or rectangular and are green with a white symbol.

Figure 1.29 An information sign

FAQ

Am I protected from electrocution if I am working on a wooden stepladder?

No, if you are working near a live current on a wooden stepladder and any metal parts of the ladder – such as the tie rods – come into contact with the current, they will conduct the flow of electricity and you may be electrocuted. Take every precaution possible in order to avoid the risk of electrocution. The simplest precaution is turning off the electricity supply.

What determines the type of scaffolding used on a job?

As you will have read in this unit, only a carded scaffolder is allowed to erect or alter scaffolding. They will select the scaffolding to be used according to the ground condition at the site, whether or not people will be working on the scaffolding, the types of materials and equipment that will be used on the scaffolding and the height to which access will be needed.

What happens if there is a delivery of timber but there is no room in the wood store?

It is probably best to remove some of the old stock from the wood store and either store it flat on timber cross-bearers or on edge in racks. This timber should be used first, and as soon as possible. The new timber should then be stored in the wood store.

What should I do if I notice a leakage in the LPG store?

Leaking LPG should be treated as a very dangerous situation. Don't turn on any lights or ignite any naked flames such as a cigarette lighter. Any kind of spark could ignite the LPG. Report the situation immediately and don't attempt to clear up the spillage yourself.

Check it out

1   What do COSHH and RIDDOR stand for?
2   Describe what might happen to you or your employer if a health and safety law is broken.
3   Write a method statement stating the actions you can take to avoid injury when lifting and carrying loads using manual handling techniques.
4   Describe the class(es) of fire that can be put out with a carbon dioxide (CO_2) extinguisher.
5   Describe why it is important to report 'near misses'.
6   State two sources of health and safety information and give a brief explanation of what services they provide.
7   Prepare a method statement describing what should be covered during a site induction.
8   State why the CSCS scheme was introduced.
9   What are the three key health and safety duties when working at height?
10 Explain the 1:4 (or 75°) ratio rule, which should be followed when erecting a ladder.

Getting ready for assessment

The information contained in this unit, as well as continued health and safety good practice throughout your training, will help you with preparing for both your end of unit test and the diploma multiple-choice test. It will also help you to understand the dangers of working in the construction industry. Wherever you work in the construction industry, you will need to understand the dangers of this occupation. You will also need to know the safe working practices for the work required for your synoptic practical assignments.

Your college or training centre should provide you with the opportunity to practise these skills, as part of preparing for the test.

You will need to know about and understand the dangers that could arise, and precautions that can be taken, for:

- the safety rules and regulations
- working at height
- knowing accident and emergency procedures
- working with electricity
- identifying hazards on site
- using personal protective equipment (PPE)
- health and hygiene
- fire and emergency procedures
- safe handling of materials and equipment
- safety signs.

You will need to apply the things you have learnt in this unit to the actual work you will be carrying out in the synoptic test, and in your professional life. For example, with learning outcome six you have seen why basic working platforms are used and the good practice you should use when working on these platforms. You have also seen the different parts of ladders and scaffolding and identified the dangers of working at height. You will now need to use this knowledge yourself when you are working, by using access equipment to the correct legislation and safeguarding you health, through using the correct PPE. You will also need to use your understanding of how PPE should be stored to maintain it in perfect condition.

Before you start work you should always think of a plan of action. You will need to know the clear sequences that materials for the practical are to be constructed to be sure you are not making mistakes as you work and that you are working safely at all times.

Your speed in carrying out these tasks in a practise setting will also help to prepare you for the time set for the test. However you must never rush the test! This is particularly important with health and safety, as you must always make sure you are working safely. Make sure throughout the test that you are wearing appropriate and correct PPE and using tools correctly.

This unit has explained the dangers you may face when working. Understanding these dangers and the precautions that can be taken to help prevent them, will not only aid you in your training, but will help you remain safe and healthy throughout your working life.

Good luck!

CHECK YOUR KNOWLEDGE

1 A risk assessment should be done:
   a when the job involves more than 50 people.
   b for every job.
   c never.
   d only when working on a scaffold.

2 Leptospirosis is also known as:
   a Weil's disease.
   b dermatitis.
   c vibration white finger.
   d none of the above.

3 The most common injury from incorrect manual handling is:
   a broken fingernails.
   b spinal injury.
   c crushing fingers under the item being handled.
   d dropping the item being handled onto toes.

4 Who is authorised to alter a scaffold?
   a Anyone.
   b The site agent.
   c A health and safety inspector.
   d A qualified, carded scaffolder.

5 What is the first thing to do if you suspect a co-worker is having an electric shock?
   a Move them away from the power source.
   b Switch off the power.
   c Dial 999.
   d Start the first aid procedure.

6 With regards to PPE, the employee must:
   a not misuse it.
   b wear it when needed.
   c report any damage to it.
   d do all of the above.

7 A fire extinguisher with a red coloured band can be used on:
   a Class A fires.
   b Class B fires.
   c Class C fires.
   d Class D fires.

8 Under RIDDOR your employer must report:
   a near misses.
   b any accident that results in the loss of three consecutive work days.
   c cut fingers.
   d all of the above.

9 When lifting manually you should:
   a have your feet shoulder-width apart.
   b have your back slightly bent.
   c keep the load away from your body.
   d lift using your back muscles.

10 Which of the following can cause electric shock?
   a Working too close to electric power lines.
   b Drilling into an internal brick/block wall.
   c Plastering walls with electric sockets.
   d All of the above.

UNIT 2002

Knowledge of information, quantities and communicating with others 2

One of the key skills in all workplaces is the ability to share knowledge and communicate effectively with the people you work with. The construction industry is no different. On any one construction project there will be a whole range of different information sources that you will need to be familiar with. These documents will affect every stage of the construction process, so any changes to them need to be communicated to everyone involved.

Construction projects often have a high budget and the work needs to be carried out to tight deadlines. Communicating clearly – both face-to-face, but also in written communications – will have a large impact on making sure a team works effectively. This unit also supports NVQ Unit VR02 Conform to Efficient Work Practices.

This unit contains material that supports the five generic units. It will also support your completion of scaled drawings throughout all TAP units.

This unit will cover the following learning outcomes:

- How to interpret and produce building information
- How to estimate quantities of resources.
- How to communicate workplace requirements effectively.

OUT

K1. Interpret and produce building information

There are a number of types of information available. You will be familiar with these from Level 1. The most commonly used are below.

Drawings

Drawings are done by the architect and are used to pass on the client's wishes to the building contractor. Drawings are usually done to scale because it would be impossible to draw a full-sized version of the project. A common scale is 1:10, which means that a line that's 10 mm long on the drawing represents 100 mm in real life. Drawings often contain symbols instead of written words to get the maximum amount of information across without cluttering the page.

These are covered in more detail on page 52–5 and include the following types of drawing:

- location drawings
- component range drawings
- assembly or detail drawings.

Work programmes

A method of showing very easily what work is being carried out when. The most common form of work programme is a bar chart, listing tasks down the left side and a timeline across the top.

Procedures

Procedures are the ways in which a company will go about doing certain tasks. Larger companies will have procedures for most things such as a procedure for ordering materials, and a procedure for making payments.

Hierarchical charts

This chart shows the level of authority and reporting lines for all people working on site, from the top (most authority) to the bottom (least authority).

Mediation

Mediation occurs after a conflict arises between two or more groups who can't agree on an outcome. A mediator is installed to listen to all sides of the debate and try to resolve the conflict by making compromises and changes so that all parties agree.

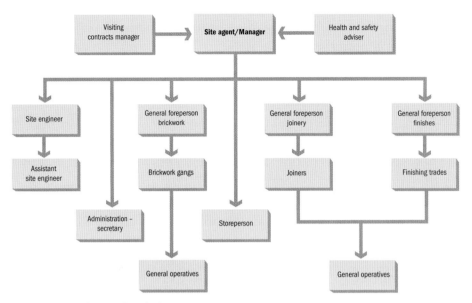

Figure 2.1 A hierarchical chart

Mediators can come from outside organisations. It is important that mediators are not seen to have any stake in the outcome of the solution. If they did, then it could influence the decision that they reach.

Before a mediation, both sides often agree to abide with the final decision of the mediator.

Disciplinary

A disciplinary is what employees receive for breaches in company rules. A disciplinary can either be verbal or written, with serious offences leading to termination of employment.

The type of disciplinary procedure used will depend on the offence. Most companies will run a three strike system wherein you will receive three warnings before you are fired.

Serious breaches such as theft or violence may result in the employee being suspended until an investigation has been carried out. If the findings state the employee has done what they are accused of, it will lead to instant dismissal.

Specifications

These accompany a drawing and give you the sizes that are not available on the drawing; it also tells you the type of material to be used and the quality that the work has to be finished to.

Manufacturer's technical information

Everything you buy, whether it is a power tool or a bag of cement, will always come with the manufacturer's technical information.

Did you know?

Organisations are required by law to give at least one written warning before dismissing an employee. However, most employees have clauses in their contract stating that they can be dismissed instantly for a serious breaches.

Remember

Reading the technical information from the manufacturers will not only give you useful information about how to use the product, it will also allow you to confirm that it is right for the job you are planning to use it for

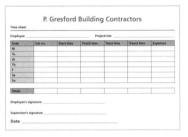

Figure 2.2 A timesheet

Figure 2.3 A day worksheet

This information will list how the component should be used and what its capabilities are.

Power tools often have their technical information provided in a booklet which will give you detailed instructions on how the machine is set up, etc.

Bagged materials such as cement, will normally have information on the bag. Even lengths of timber will have technical information, but for this you may need to contact the manufacturer. With lengths of timber this is important as timber is stress graded and you need to be sure that the materials you are planning to use are up for the job.

Organisation documentation

No building site functions without paperwork. Some of the key documents include:

- **timesheets** – these record hours worked and are completed by every employee individually. They can be used to work out how many hours the client will be charged for
- **day worksheets** – often confused with timesheets, but different as they are used when there is no price or estimate for the work, to enable the contractor to charge. They record work done, hours worked and sometimes materials used
- **job sheets** – these are used when the work has already been priced and enable the worker to see what needs to be done and the site agent to see what has been completed
- **variation orders** – used by the architect to make any changes to the original plans, including omissions, alterations and extra works
- **confirmation notices** – given to the contractor to confirm any changes made in the variation order, so that the contractor can go ahead.

P. Gresford Building Contractors

Job sheet

Customer	Chris MacFarlane
Address	1 High Street
	Any Town
	Any County

Work to be carried out

Hang internal door in kitchen

Special conditions/Instructions

Fit with door closer

3 x 75 mm butt hinges

Figure 2.4 A job sheet

VARIATION TO PROPOSED WORKS AT 123 A STREET

REFERENCE NO:

DATE _____

FROM _____

TO _____

POSSIBLE VARIATIONS TO WORK AT 123 A STREET

| ADDITIONS |
| |
| |
| OMISSIONS |
| |
| |

SIGNED _____

Figure 2.5 A variation order

CONFIRMATION FOR VARIATION TO PROPOSED WORKS AT 123 A STREET

REFERENCE NO:

DATE _____

FROM _____

TO _____

I CONFIRM THAT I HAVE RECEIVED WRITTEN INSTRUCTIONS FROM _____
POSITION _____
TO CARRY OUT THE FOLLOWING POSSIBLE VARIATIONS TO THE ABOVE NAMED CONTRACT

| ADDITIONS |
| |
| |
| OMISSIONS |
| |

SIGNED _____

Figure 2.6 A confirmation notice

Figure 2.7 A requisition form

Figure 2.8 A delivery note

Figure 2.9 An invoice

- **orders/requisitions** – used to order materials from a supplier
- **delivery notes** – given to the contractor by the supplier, listing all the materials and components being delivered. Each should be checked for accuracy against the order and the delivery (to ensure what is delivered is correct and that it matches the note)
- **invoices** – these come from a variety of sources and state what services or goods have been provided and the charge for it
- **delivery records** – these list all deliveries over a certain period (usually a month) and are sent to the contractor's Head Office so that payment can be made
- **daily report** – used to pass general information on to a company's Head Office
- **accidents and near-miss reports** – a legal requirement; a company has an accident book in which all reports must be made. Reports must also be made when an accident nearly happened but did not occur – a 'near miss'.

Figure 2.10 A delivery record

Figure 2.11 A daily report

Figure 2.12 An accident/near-miss report

Training and development records

As learners in the construction industry you are currently doing training that will develop your skills. During your training there will be records of what you have been trained in. These are used at the end as evidence so that you can achieve your qualification.

However, training doesn't stop as soon as you qualify! Further training and development is important in the construction industry, especially with technological advances in tools, methods of working, etc. There will be certain tasks you are introduced to at work that will require you to receive some training to carry them out, e.g. any employee who uses tools such as nail guns must be trained in their uses before they can use them. Even ensuring that all employees are CSCS carded will require the training and development of all employees.

Records of all training must be kept, as if you leave one employer you may need to prove that you have received training before another employer will hire you.

Checking information for conformity

As with all documents, the information above needs to be checked for **conformity**. Using documents that don't conform to, or meet, the company's standards could cause problems, delays or confusion in the building process. For example, faxing a blank piece of paper with a few materials on it to a supplier may be rejected by suppliers which will lead to materials not being ordered and delays to the build.

Only documents that have been approved must be used. If in doubt ask your supervisor.

Contract documents and interpreting specifications

Contract documents are vital to a construction project. They are created by a team of specialists – the architect, structural engineer, services engineer and quantity surveyor – who first look at the draft of drawings from the architect and client. The contract documents this team goes on to produce will vary depending on the size and type of work being done, but will usually include:

- plans and drawings
- specification

- schedules
- bill of quantities
- conditions of contract.

Plans and drawings have already been covered, so we will start with the specification.

Specification

The specification or 'spec' is a document produced alongside the plans and drawings and is used to show information that cannot be shown on the drawings. Specifications are almost always used for things such as:

- foundations
- walls
- materials
- surface finish
- floors
- roofs
- components.

The only exceptions might be in the case of very small contracts. A specification should contain:

- **site description** – a brief description of the site including the address
- **restrictions** – what restrictions apply such as working hours or limited access
- **services** – what services are available, what services need to be connected and what type of connection should be used
- **materials description** – including type, size, quality, moisture content, etc.
- **workmanship** – including methods of fixing, quality of work and finish.

The specification may also name subcontractors or suppliers, or give details such as how the site should be cleared.

Schedules

A schedule is used to record repeated design information that applies to a range of components or fittings. Schedules are mainly used on bigger sites where there are multiples of several types of house (4-bedroom, 3-bedroom, 3-bedroom with dormers, etc.), each type having different components and fittings. The schedule prevents the wrong component or fitting being put in the wrong house. Schedules can also be used on smaller jobs such

Figure 2.13 A good 'spec' helps avoid confusion when dealing with subcontractors or suppliers

as a block of flats with 200 windows, where there are six different types of window.

Bill of quantities

The bill of quantities is produced by the quantity surveyor. It gives a complete description of everything that is required to do the job, including labour, materials and any items or components, based on information from the drawings, specification and schedule. The same bill of quantities is sent out to all prospective contractors so they can submit a tender based on the information – this helps the client select the best contractor for the job.

Figure 2.14 Every item needed should be listed on the bill of quantities

All bills of quantities contain the following information:

- **preliminaries** – general information such as the names of the client and architect, details of the work and descriptions of the site
- **preambles** – similar to the specification, outlining the quality and description of materials and workmanship, etc.
- **measured quantities** – a description of how each task or material is measured with measurements in metres (linear and square), hours, litres, kilograms or simply the number of components required
- **provisional quantities** – approximate amounts where items or components cannot be measured accurately
- **cost** – the amount of money that will be charged per unit of quantity.

The bill of quantities may also contain:

- any costs that may result from using subcontractors or specialists
- a sum of money for work that has not been finally detailed
- a sum of money to cover contingencies for unforeseen work.

Item ref no	Description	Quantity	Unit	Rate £	Cost £
A1	Treated 50 × 225 mm sawn carcass	200	M		
A2	Treated 75 × 225 mm sawn carcass	50	M		
B1	50 mm galvanised steel joist hangers	20	N/A		
B2	75 mm galvanised steel joist hangers	7	N/A		
C1	Supply and fit the above floor joists as described in the preambles				

Figure 2.15 Extract from a bill of quantities

This is an extract from a bill of quantities that might be sent to prospective contractors, who would then complete the cost section and return it as their tender.

To ensure that all contractors interpret and understand the bill of quantities, the Royal Institution of Chartered Surveyors and the Building Employers' Confederation produce a document called the Standard Method of Measurement of Building Works (SMM). This provides a uniform basis for measuring building work, for example stating that carcassing timber is measured by the metre whereas plasterboard is measured in square metres.

Conditions of contract

Almost all building work is carried out under a contract. A small job with a single client (e.g. a loft conversion) will have a basic contract stating that the contractor will do the work to the client's satisfaction, and that the client will pay the contractor the agreed sum of money once the work is finished. Larger contracts with clients such as the Government will have additional clauses, terms or stipulations, which may include any of the below.

Variations

A variation is a modification of the original drawing or specification. The architect or client must give the contractor written confirmation of the variation, then the contractor submits a price for the variation to the quantity surveyor (or client, on a small job). Once the price is accepted, the variation work can be completed.

Interim payment

An **interim** payment schedule may be written into the contract, meaning that the client pays for the work in instalments. The client may pay an amount each month, linked to how far the job has progressed, or may make regular payments regardless of how far the job has progressed.

Final payment

Here the client makes a one-off full payment once the job has been completed to the specification. A final payment scheme may also have additional clauses included, e.g.:

- **retention** – this is when the client holds a small percentage of the full payment back for a specified period (usually six months). It may take some time for any defects to show such as cracks in plaster. If the contractor fixes the defects, they will receive the retention payment; if they don't fix them, the retention payment can be used to hire another contractor to do so

> **Did you know?**
>
> On a poorly run contract, a penalty clause can be very costly and could incur a substantial payment. In an extreme case, the contractor may end up making a loss instead of a profit on the project

Figure 2.16 Maintaining a good relationship will keep the job running smoothly

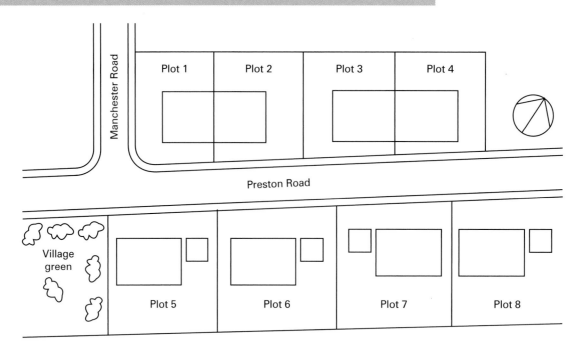

Figure 2.17 Block plan showing location

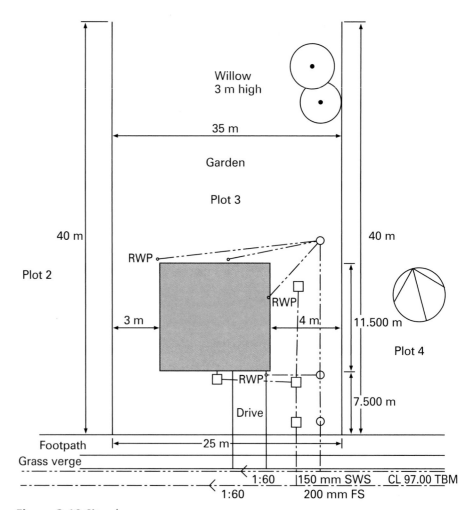

Figure 2.18 Site plan

Functional skills

FM 1.2.1b relates to interpreting information from sources such as diagrams, tables, charts and graphs. Examples are shown on pages 52–55. You will be practising both these functional skills when you use contract documents and plans.

- **penalty clause** – this is usually introduced in contracts with a tight deadline, where the building must be finished and ready to operate on time. If the project overruns, the client will be unable to trade in the premises and will lose money, so the contractor will have to compensate the client for lost revenue.

Types of drawing

Working drawings

Working drawings are scale drawings showing plans, elevations, sections, details and location of a proposed construction. They can be classified as:

- location drawings
- component range drawings
- assembly or detail drawings.

Location drawings

Location drawings include block plans and site plans.

Block plans identify the proposed site by giving a bird's eye view of the site in relation to the surrounding area. An example is shown in Figure 2.17.

Site plans give the position of the proposed building and the general layout of the roads, services, drainage, etc. on site. An example is shown in Figure 2.18.

Component range drawings

Component range drawings show the basic sizes and reference system of a standard range of components produced by a manufacturer. This helps in selecting components suitable for a task and available off-the-shelf. An example is shown in Figure 2.19.

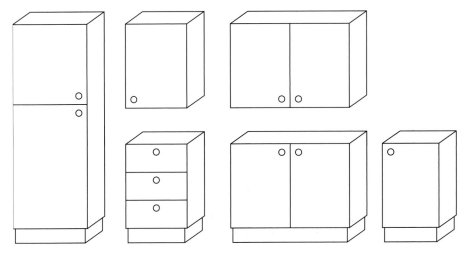

Figure 2.19 Component range drawing

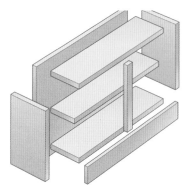

Figure 2.20 Assembly drawing

Assembly or detail drawings

Assembly or detail drawings give all the information required to manufacture a given component. They show how things are put together and what the finished item will look like. An example is shown in Figure 2.20.

Types of projection

Building, engineering and similar drawings aim to give as much information as possible in a way that is easy to understand. They frequently combine several views on a single drawing.

These may be elevations (the view we would see if we stood in front or to the side of the finished building) or plan (the view we would have if we were looking down on it). The view we see depends on where we are looking from. There are different ways of 'projecting' what we would see onto the drawings.

The two main methods of projection, used on standard building drawings, are orthographic and isometric.

Orthographic projection

Orthographic projection works as if parallel lines were drawn from every point on a model of the building on a sheet of paper held up behind it (an elevation view), or laid out underneath it (plan view).

There are different ways that we can display the views on a drawing. The method most commonly used in the building industry for detailed construction drawings, is called 'third angle projection'. In this the front elevation is roughly central. The plan view is drawn directly below the front elevation and all other elevations are drawn in line with the front elevation. An example is shown in Figure 2.21.

Front elevation

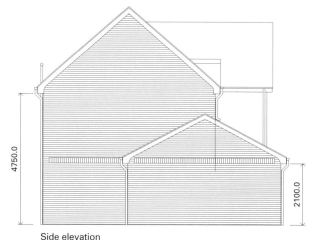

Side elevation

Figure 2.21 Orthographic projection

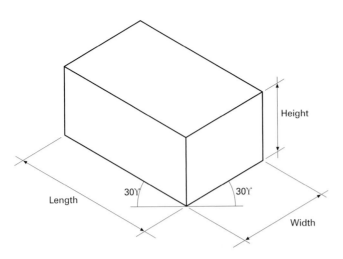

Figure 2.22 Isometric projection of rectangular box

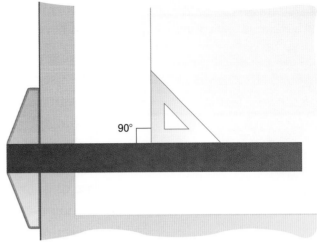

Figure 2.23 Drawing vertical lines

Isometric projection

In isometric views, the object is drawn at an angle where one corner of the object is closer to the viewer. Vertical lines remain vertical but horizontal lines are drawn at an angle of 30° to the horizontal. This can be seen in Figure 2.22, which shows a simple rectangular box.

Figures 2.23 and 2.24 show the method of drawing these using a T-square and set square.

Datum points

The need to apply levels is required at the beginning of the construction process and continues right up to the completion of building. The whole country is mapped in detail and the Ordnance Survey place datum points (bench marks) at suitable locations from which all levels can be taken.

The site datum gives a reference point on site from which all levels can be related.

Title panels

Every drawing must have a title panel, which is normally located in the bottom right-hand corner of each drawing sheet. See Figure 2.25 for an example.

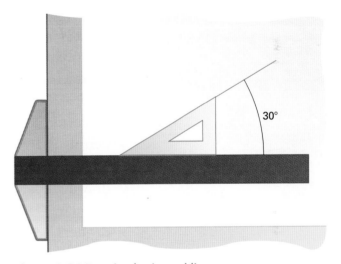

Figure 2.24 Drawing horizontal lines

ARCHITECTS	**CLIENT**
Peterson, Thompson Associates	Carillion Development
237 Cumberland Way	
Ipswich	**JOB TITLE**
IP3 7FT	Appleford Drive
	Felixstowe
Tel: 01234 567891	4 bed detached
Fax: 09876 543210	
Email: enquiries@pta.co.uk	
	SCALE 1:50
DRAWING TITLE	
Plan – garage	**DRAWING NO:** 2205-06
DATE: 27.08.2008	**DRAWN BY:** RW

Figure 2.25 Typical title panel

The information contained in the panel is relevant to that drawing only and contains the following:

- drawing title
- scale used
- draughtsman's name
- drawing/project number
- company name
- job/project title
- date of drawing
- revision notes
- projection type.

Drawing equipment

A set of good quality drawing equipment is required when producing drawings. It should include:

- set squares (A)
- protractors (B)
- compasses (C)
- scale rule (D)
- pencils (E)
- eraser (F)
- drawing board (G)
- T-square (H)
- dividers.

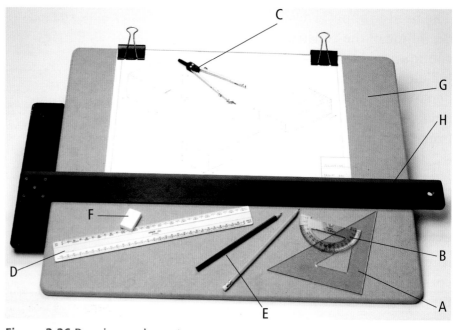

Figure 2.26 Drawing equipment

Set squares

Two set squares are required, a 45° set square and a 60°/30° set square. These are used to draw vertical and inclined lines. A 45° set square (A) is shown in the photograph.

Protractors

Protractors (B) are used for setting out and measuring angles.

Compasses and dividers

Compasses (C) are used to draw circles and arcs. Dividers (not shown) are used for transferring measurements and dividing lines.

Scale rule

A scale rule that contains the following scales is recommended:

1:5/1:50   1:10/1:100   1:20/1:200   1:250/1:2500

An example (D) is shown in the photo.

The main scales that may be used on location drawings are:

- 1:1
- 1:2
- 1:5
- 1:10
- 1:20
- 1:50

Pencils

Two pencils (E) are required:

- HB for printing and sketching
- 2H or 3H for drawing.

Eraser

A vinyl or rubber eraser (F) is required for alterations or corrections to pencil lines.

Drawing board

A drawing board (G) is made from a smooth, flat surface material, with square and parallel edges.

T-square

The T-square (H) is used mainly for drawing horizontal lines.

> **Did you know?**
> Set squares, protractors and rules should occasionally be washed in warm, soapy water

K2. Estimate quantities of resources

To complete estimates of calculations, you will need to be familiar with the mathematical principles behind these calculations.

Calculations

Throughout your career in the construction industry you will have to make use of numbers and calculations in order to plan and carry out work. You will therefore need to make sure you are confident dealing with numbers, which may mean that you have to develop and improve your maths and numeracy skills.

To make estimates, you will need to be familiar with some calculation techniques used in the industry. These methods will help you to calculate the amounts of materials you will need to use on a project.

Decimal numbers

Most of the time, the numbers we use are whole numbers. For example, we might buy six apples, or two loaves of bread or one car. However, sometimes we need to use numbers that are less than whole numbers, for example, we might eat one and a quarter sandwiches, two and a half cakes and drink three-quarters of a cup

Hundred	Tens	Units	.	Tenths	Hundredths	Thousandth	
		1	.	8			1.8 = 1 unit and 8 tenths 1 whole $\frac{8}{10}$
	5	6	.	4	5		56 and 45 hundredths 45 hundredths
2	9	0	.	0	1	7	290 and 17 thousandths

Table 2.1 A place value table for digits to the right of a decimal point

of tea. You can use decimals to show fractions or parts of quantities. Table 2.1 shows the value of the digits to the right of a decimal point.

Rounding to a number of decimal places

To round a number to a given number of decimal places, look at the digit in the place value position after the one you want.

- If it is 5 or more, round up.
- If it is less than 5, round down.

Say we wanted to round 4.634 to two decimal places, the digit in the third decimal place is 4, so we round down. Therefore, 4.634 rounded to 2 decimal places (d.p.) is 4.63.

If we look at the number 16.127, the digit in the third decimal place is 7, so we round up. Therefore, 16.127 rounded to 2 d.p. is 16.13.

Rounding to a number of significant figures

The most significant figure in a number is the digit with the highest place value. To round a number to a given number of significant figures, look at the digit in the place value position after the one you want.

- If it is 5 or more, round up.
- If it is less than 5, round down.

Say we wanted to write 80,597 to one significant figure, the most significant figure is 8. The second significant figure is 0, so we round down. Therefore, 80,597 to 1 significant figure (s.f.) is 80,000.

If we wanted to write 80,597 to two significant figures, the first two significant figures are 8 and 0. The third significant figure is 5, so we round up. Therefore, 80,597 to 2 s.f. is 81,000.

Multiplying and dividing by 10, 100, 1000...

- To multiply a number by 10, move the digits one place value to the left.
- To multiply a number by 100, move the digits two place values to the left.
- To multiply a number by 1000, move the digits three place values to the left.

For example:

```
                5  3.25 × 10 =        3.25
5 × 10 =       50  3.25 × 10 =        32.5
5 × 100 =     500  3.25 × 100 =      325
5 × 1000 =   5000  3.25 × 1000 =   3250
```

Did you know?

Knowing about place value helps you to read numbers and to put numbers and quantities in order of size

Remember

If a calculation results in an answer with many decimal places such as 34.568923, you can round to 1 or 2 decimal places to make it simpler

Did you know?

Knowing how to multiply and divide by 10, 100, 1000, etc. is useful for converting metric units of measurement and finding percentages

- To divide a number by 10, move the digits one place value to the right.
- To divide a number by 100, move the digits two place values to the right.
- To divide a number by 1000, move the digits three place values to the right.

For example:

$$80\,000 \div 10 = 8000 \qquad 473.6 \div 10 = 47.36$$
$$80\,000 \div 100 = 800 \qquad 473.6 \div 100 = 4.736$$
$$80\,000 \div 1000 = 80 \qquad 473.6 \div 1000 = 0.4736$$

Converting decimals to fractions

You can use place value to convert a decimal to a fraction. For example:

0.3 is 3 tenths which is $\frac{3}{10}$.

0.25 is 25 hundredths which is $\frac{25}{100}$.

$\frac{25}{100}$ simplifies to $\frac{1}{4}$ (by dividing the top and bottom numbers by 25).

Table 2.x shows some useful fraction/decimal equivalents.

See page 62 for more on simplifying fractions.

Did you know?

Knowing how to convert between fractions and decimals helps with working out parts of quantities and calculating percentages

Multiples

Multiples are the numbers you get when you multiply any number by other numbers, e.g.:

- the multiples of 3 are 3, 6, 9, 12, 15, 18, 21, 24, 27, 30…
- the multiples of 4 are 4, 8, 12, 16, 20, 24, 28, 32, 36, 40…
- the multiples of 5 are 5, 10, 15, 20, 25, 30, 35, 40, 45, 50…

Common multiples

See the multiples of 3 and 5 below.

Multiples of 3: 3, 6, 9, 12, 15, 18, 21, 24, 27, 30, 33, 36…

Multiples of 5: 5, 10, 15, 20, 25, 30, 35…

3 and 5 have the multiples 15 and 30 in common. 15 and 30 are common multiples of 3 and 5. The lowest common multiple of 3 and 5 is 15.

Remember

The multiples of a number are the numbers in its 'times' (multiplication) table

Did you know?

Knowing how to find the lowest common multiple of two numbers helps in adding and subtracting fractions

Factors

The factors of a number are the whole numbers that divide into it exactly, e.g.:

- the factors of 18 are 1, 2, 3, 6, 9 and 18
- the factors of 30 are 1, 2, 3, 5, 6, 10, 15 and 30
- 5 is a factor of 5, 10, 15, 20…
- 7 is a factor of 7, 14, 21, 28…

Did you know?

Knowing multiplication tables will help you to work out factors

Common factors

See the factors of 28 and 36 below.

- The factors of 28 are 1, 2, 4, 7, 14, 28
- The factors of 36 are 1, 2, 3, 4, 6, 9, 12, 18, 36

From these you can see that 28 and 36 have the factors 1, 2 and 4 in common. 1, 2 and 4 are the common factors of 28 and 36. 4 is the highest common factor of 28 and 36.

Did you know?

Knowing how to find common factors of numbers helps with simplifying fractions

Fractions

Fractions describe parts of a whole, for example a half of a pie, a third of a can of cola or a quarter of a cake.

In a fraction:

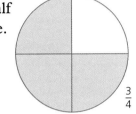

$\frac{3}{4}$

$\dfrac{3}{4}$ the top number is called the **numerator**

the bottom number is called the **denominator**

The denominator shows how many equal parts the whole is divided into. The numerator shows how many of those parts you have.

Finding a fraction of a quantity

To find a fraction of a quantity, you divide by the denominator and multiply by the numerator, e.g.:

- to find $\frac{1}{2}$ of 500 m, divide by 2 (500 m ÷ 2 = 250 m)
- to find $\frac{2}{5}$ of £40, divide by 5 (£40 ÷ 5 = £8) and multiply by 2 (£8 × 2 = £16).

Did you know?

To find equivalent fractions you can:

- multiply the numerator and denominator by the same number
- divide the numerator and denominator by the same number

Equivalent fractions

Two fractions are equivalent if they have the same value, e.g.:

$\dfrac{1}{2} = \dfrac{2}{4} = \dfrac{3}{6} = \dfrac{4}{8}$

 $\frac{1}{2}$ $\frac{2}{4}$ $\frac{3}{6}$ 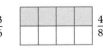 $\frac{4}{8}$

$\dfrac{2}{3} = \dfrac{4}{6} = \dfrac{6}{9} = \dfrac{8}{12}$

 $\frac{2}{3}$ $\frac{4}{6}$ $\frac{6}{9}$ $\frac{8}{12}$

Unit 2002 Knowledge of information, quantities and communicating with others 2

Did you know?

To simplify a fraction, divide the numerator and denominator by a common factor of both

Simplifying fractions

To simplify a fraction, write it as an equivalent fraction with smaller numbers in the numerator and denominator.

For example, $\frac{8}{12}$ simplifies to $\frac{2}{3}$.

When a fraction cannot be simplified any more, it is in its simplest form, or its lowest terms.

For example:

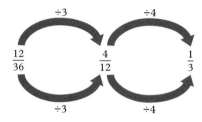

$\frac{12}{36} = \frac{1}{3}$ in its simplest form.

Multiplying fractions

To multiply a fraction by a whole number, multiply the numerator by the whole number.

For example:

$\frac{2}{9} \times 4 = \frac{8}{9}$

To multiply a fraction by another fraction, multiply the numerators and the denominators, e.g.:

$\frac{2}{3} \times \frac{5}{8} \times \frac{2 \times 5}{3 \times 8} = \frac{10}{24}$

Give the answer in its simplest form:

$\frac{10}{24} = \frac{5}{12}$ (dividing numerator and denominator by 2)

Dividing fractions

To divide one fraction by another, you invert the fraction you are dividing by (i.e. turn it upside down), and multiply.

For example:

$\frac{3}{4} \div \frac{2}{3} = \frac{3}{4} \times \frac{3}{2} = \frac{9}{8} = 1\frac{1}{8}$

To divide a fraction by a whole number, or to divide a whole number by a fraction, write the whole number as a fraction with denominator 1, and use the same method.

$\frac{4}{5} \div 3 = \frac{4}{5} - \frac{3}{1} = \frac{4}{5} \times \frac{1}{3} = \frac{4}{15}$

$1 \div \frac{3}{4} = \frac{6}{1} \div \frac{3}{4} = \frac{6}{1} \times \frac{4}{3} = \frac{24}{3} = 8$

Adding fractions

To add fractions with the same denominator, add the numerators.

For example:

$\frac{1}{3} + \frac{1}{3} = \frac{2}{3}$

$\frac{1}{5} + \frac{3}{5} = \frac{4}{5}$

To add fractions with different denominators, first write the fractions as equivalent fractions with the same denominator. Use the lowest common multiple of the two denominators. You can use any common multiple as the denominator, but using the lowest common multiple keeps the numbers smaller and the calculations simpler.

For example:

$\frac{1}{2} + \frac{1}{3}$

The denominators are not the same. The lowest common multiple of 2 and 3 is 6.

To write $\frac{1}{2}$ as an equivalent fraction with denominator 6, multiply numerator and denominator by 3, to give $\frac{3}{6}$.

To write $\frac{1}{3}$ as an equivalent fraction with denominator 6, multiply numerator and denominator by 2, to give $\frac{2}{6}$.

The calculation is $\frac{3}{6} + \frac{2}{6} = \frac{5}{6}$.

A mixed number has a whole number and a fraction part, for example $3\frac{1}{4}$.

To add mixed numbers, add together the whole number parts and then the fractions.

For example, if we wanted to add $1\frac{1}{2}$ and $2\frac{1}{3}$:

Add the whole numbers: $1 + 2 = 3$.

Now add the fractions: $\frac{1}{2} + \frac{1}{3} = \frac{3}{6} + \frac{2}{6} = \frac{5}{6}$.

Combine the two answers: $1\frac{1}{2} + 2\frac{1}{3} = 3\frac{5}{6}$.

Subtracting fractions

To subtract fractions with the same denominator, subtract the numerators.

For example:

$$\frac{7}{8} - \frac{3}{8} = \frac{4}{8} = \frac{1}{2}$$

$$\frac{7}{8} \qquad \frac{3}{8} \qquad \frac{4}{8}$$

To subtract mixed numbers, first write them as improper (top heavy) fractions with a common denominator.

For example: $2\frac{1}{3} - \frac{1}{2} = \frac{7}{3} - \frac{1}{2} = \frac{14}{6} = 5\frac{5}{6}$

Percentages

Percentages are another way of showing parts of a quantity. Percentage means 'number of parts per hundred'. The symbol % means per cent, e.g.:

1% means 1 out of a hundred or $\frac{1}{100}$

10% means 10 out of a hundred or $\frac{10}{100}$

84% means 84 out of a hundred or $\frac{84}{100}$

100% means the whole quantity.

10%

84%

Finding a percentage of a quantity

To find a percentage of a quantity, find 1% first by dividing by 100, then multiply by the number you need, e.g.:

20% of £45

1% of £45 = $\frac{£15}{100}$ = £45 ÷ 100 = £0.45

So 20% of £45 = 20 × £0.45 = £9

Percentage change

A number can be increased or decreased by a percentage. Wages are often increased by a percentage (e.g. a 4% rise in wages). Items are often reduced by a percentage in sales (e.g. 10% off or a 20% reduction).

Did you know?

Percentages are also used for:

- paying deposits – a deposit is a percentage of the whole price (e.g. 20% deposit)
- paying interest – interest is a percentage of money. It is repaid on top of the money (e.g. a loan from a bank of £1000 at 15% interest)
- profit – when charging a client for work you have carried out, you will need to add on to your costs a percentage for your profit
- Value Added Tax (VAT) – VAT is a government tax added to many items or goods that we buy (the standard VAT rate is 17.5%)

Example

A set of paintbrushes costs £12.99. In the sale they are 15% off.

1. How much money do you save by buying the paintbrushes in the sale?

2. What is the sale price of the paintbrushes?
   Work out 15% of £12.99
   1% of £12.99 = £0.1299
   So 15% of £12.99 = 15 × £0.1299 = £1.9485 = £1.95 to the nearest penny
   You save £1.95

3. The sale price is
   £12.99 – £1.95 = £11.04

Ratio

A **ratio** describes a relationship between quantities. You can read a ratio as a 'for every' statement, e.g.:

Green paint is made by mixing blue and yellow in the ratio 1:2. This means, for every 1 litre of blue paint you need 2 litres of yellow paint.

A labourer and a bricklayer agree to share their bonus in the ratio 2:5. The bonus is £42. The ratio 2:5 means that for every 2 parts the labourer receives, the bricklayer receives 5 parts. Thus, the bonus needs to be split into 2 + 5 = 7 parts. One part can be calculated: £42 ÷ 7 = £6. Therefore, the labourer receives two parts = 2 × £6 = £12 and the bricklayer receives five parts = 5 × £6 = £30.

Volume

Volume is the amount of space taken up by a 3-D or solid shape. Volume is measured in cube units such as cubic centimetres (cm³) and cubic metres (m³).

A cuboid is a 3-D shape whose faces are all rectangles.

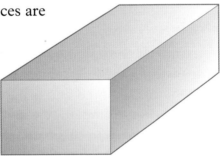

A cube is a 3-D shape whose faces are all squares.

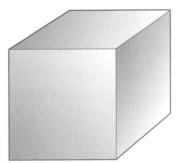

This cuboid is made of 1cm³ cubes.

You can find the volume by counting the cubes. Volume = number of cubes = 36 cm³.

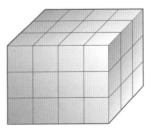

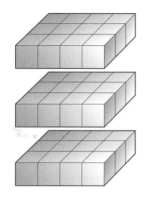

You can also calculate the volume by splitting the solid into equal rectangular layers.

Each layer has 4 × 3 cubes.

There are three layers, so the total number of cubes is
3 × 4 × 3 = 36 cm³.

The volume of a cuboid with length *l*, width *w* and height *h* is

$$V = l \times w \times h$$

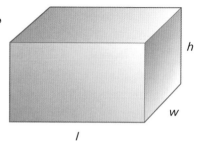

For a cube, length = width = height, so the volume of a cube with side *l* is:

$$V = l3$$

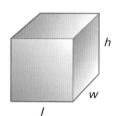

You can find the volume of concrete needed for a rectangular floor by measuring the length and width of the floor, and the depth of the concrete required and using the formula for the volume of a cuboid. For example, we can work out the volume of concrete needed for the floor of a rectangular room with length 3.7 m and width 2.9 m, if the depth of the concrete is 0.15 m.

Visualise the floor as a cuboid.

The depth of the concrete (0.15 m) is the height of the cuboid.

$$V = l \times w \times h$$
$$= 3.7 \times 2.9 \times 0.15$$
$$= 1.6095 \text{ m}^3$$
$$= 1.61 \text{ m}^3 \text{ (to 2 d.p.)}$$

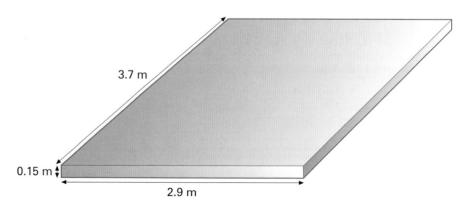

Units of volume

Volume is measured in **cube units** such as mm³, cm³, m³. The volume of this cube is 1 cm³ or 10 × 10 × 10 = 1000 mm³.

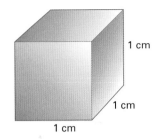

1 cm
1 cm
1 cm

The volume of this cube is 1 m3 or 100 × 100 × 100 = 1 000 000 cm³.

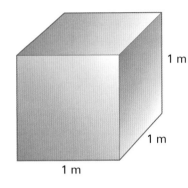

1 m
1 m
1 m

> **Remember**
>
> 1 cm³ = 1000 mm³
> 1 m³ = 1 000 000 cm³

A cuboid is a 3-D shape with rectangular faces (like a box). The formula for the volume of a cuboid is $V = l \times w \times h$. For example, we can calculate the volume of this cuboid:

1. in cm³
2. in m³
3. $V = l \times w \times h = 56 \times 84 \times 221 = 1\ 039\ 584$ cm³
4. $1\ 039\ 584 \div 1\ 000\ 000 = 1.039584$ m² = 1.04 m³ (to 2 d.p.)

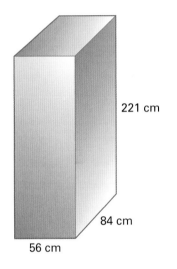

221 cm
84 cm
56 cm

Rounding to a sensible degree of accuracy

Sometimes measurement calculations give an answer to a large number of decimal places. It is sensible to round the answer to a measurement that is practical. For example, Ahmed has a piece of wood that is 190 cm long. He wants to cut it into 7 equal lengths.

> **Remember**
>
> In some situations it is most sensible to round up

He works out 190 ÷ 7 = 14.28578142 cm.

You cannot measure 0.28578142 of a centimetre!

It is sensible to round to 14.3 cm (to 1 d.p.), which is 14 cm and 3 mm can be measured.

Example

A plasterer calculates the total area of walls in a room as 36 m².

Plasterboard sheets have an area of 2.88 m².

The number of plasterboard sheets needed is
$36 \div 2.88 = 12.5$

If the plasterer buys 12 sheets he will not have enough

If he buys 13 sheets he will have half a sheet left over

In this case it is sensible to round up, and buy 13 sheets

You may find there are times when you need to round down.

In some situations it is most sensible to round down. How many 2 metre lengths can be cut from 7 metres of pipe?

$7 \div 2 = 3.5$

You can cut three 2 metre lengths. The rest (0.5 of a 2 metre length) is wasted.

In this case it is sensible to round down: you can only cut three 2 metre lengths.

Using calculations

There are three main areas where you will need to use these types of calculation in a construction setting.

The tender process

Tendering is a competitive process where the contractor works with a specification and drawings from the client and submits a cost estimate for the work (including materials, labour and equipment). Tenders are often invited for large contracts such as Government contracts, with strict fixed deadlines for the tenders to be received.

An estimator will calculate the total cost in the tender. Using the information in the specification the estimator calculates the amount of materials and labour needed to complete the work. The final tender is based on this estimation.

All the tenders for a contract will then present their case and costs to the client, who will then decide on one business to be offered the contract.

Quoting

A quote is basically part of the tender process but it will only contain pricing information on materials, labour, etc. The quote will state how much the job will cost without any additional information that may appear on a tender such as using a percentage of the local work force or recycling a certain amount of materials.

The quote is then used as part of the tender to give an idea of the potential cost of a job. Companies submitting tenders will look to make this quote as attractive as possible to the client.

Estimated pricing

Estimated pricing is used to create the quote. An estimator will look at what is required and provide an estimated price for it.

Tenders for jobs may take many months for the successful tender to be selected so an estimator who prices everything up exactly as it is now may be wrong in six months time as the price of labour or materials may have changed. This means they will give an estimated price instead, based on a calculation of how much the materials or labour may cost in the future.

The resources used for making an estimated price and then a final price include:

- materials
- purchase orders
- invoices
- basic time study sheets
- labour schedules
- information technology
- job sheets
- site diaries
- equipment availability lead times
- building supplier's price lists
- book systems used for pricing.

Predicting waste

When working out what materials you require to complete a job it is advisable to add on a certain amount to your calculations for material waste.

This is necessary because if you order the exact amount you require you are not considering any offcuts or damages to the material. When cementing or plastering, not all the plaster will go on the wall. Some will fall on the floor and some may go off too quickly if you mix up too much to use in one day. It is generally accepted that between 5% and 10% is added on to the total to allow for waste.

For adding on a percentage you simply need to divide the amount by 100 and then multiply by 100 + whatever the percent is.

Deciding what materials to use

The materials used for the job will depend solely on what the architect agrees with the client. Larger companies will have an agreement with suppliers so that they can order all their materials with them at a discounted rate.

Any specialist materials will be resourced by a buyer. They will look at which companies provide the materials, what the cost is

Remember

You need to make allowance for mistakes, such as cutting materials wrong, as well as natural waste

Example

300 metres of timber is the required amount and you want to add on 7% for waste so:

$\frac{300}{100} = 3$

$3 \times 107 = 321$

With 7% waste you would order 321 metres

and what attributes the company has, e.g. do they work with fair trade, etc.

A small company on a smaller job will discuss this with the client. The client may wish to order the materials themselves so that they are sure they are getting exactly what they want. If not, the company will get the materials from suppliers such as a local builder's merchant.

Estimating labour rates

The way that labour is paid can be split into two different ways.

Day work/hourly rate

This rate is used when the tradesperson will be paid a specific amount for every hour that they work. The amount will depend on where the work is being carried out as the cost of living is different in each area.

Places where the cost of living is low may receive £10–£20 per hour. In areas with a high cost of living (such as London) the rate may be £20–£30 per hour. The experience of a worker will also affect the day work rate. Newly qualified Level 2 apprentices will not be paid the same as someone with 30 years of experience.

Price work

This rate is used when the tradesperson will be paid for the work they carry out. Examples of this include a carpenter who receives £20 for every door they hang or a painter who gets £300 for every flat they decorate. This method is often preferred, though it means that you may have to work harder. However, the more work you do, the more you will earn.

Again, the prices for these will vary not only from area to area but also within trades. A carpenter may get paid £2000 to fit a truss roof but only £250 to fit a small kitchen in a flat. A painter may get paid £15 to paint the inside of a window compared with £20 to paint the outside. These differences in price are worked out prior to the job starting and take into account things such as weather or hazards. The roof may look like the best job at £2000, but if it is raining heavily for a week, or alterations to the scaffold are required, you may not get much work done. You may be able to fit seven kitchens in a week no matter what the weather.

The price work method is calculated by working out how many hours it will take to complete the task and then giving a certain price based on the day work rate.

Did you know?

When estimating the labour costs for a job it is easier to use the day work rate method as you can calculate how many hours the job will take

For example, the day work rate may be £20 per hour and a roof should take 100 hours. This means a price of £2000 will be put forward.

Remember

The £2000 price is for the whole roof. If four people work on it they will get paid £500 each, not the full £2000

K3. Communicate workplace requirements effectively

Within the construction industry there is a range of different job and career types that you could choose from. What all these job roles have in common, however, is that clear communication with co-workers is always vital.

Job and careers

Jobs and careers in the construction industry fall mainly into one of four categories:

- building
- civil engineering
- electrical engineering
- mechanical engineering.

Building involves the physical construction (making) of a structure. It also involves the maintenance, restoration and refurbishment of structures.

Civil engineering involves the construction and maintenance of work such as roads, railways and bridges.

Electrical engineering involves the installation and maintenance of electrical systems and devices such as lights, power sockets and electrical appliances.

Mechanical engineering involves the installation and maintenance of things such as heating, ventilation and lifts.

The category that is the most relevant to your course is building.

Job types

The construction industry employs people in four specific areas:

1. professionals
2. technicians
3. building craft workers
4. building operatives.

Professionals

Professionals are generally of graduate level (i.e. people who have a degree from a university) and may have one of the following types of job in the construction industry:

- architect – someone who designs and draws the building or structure
- structural engineer – someone who oversees the strength and structure of the building
- land surveyor – someone who checks the land for suitability to build on
- building surveyor – someone who provides advice on construction projects
- service engineer – someone who plans the services needed within the building, for example gas, electricity and water supplies.

Technicians

Technicians link professional workers with craft workers and are made up of the following people:

- architectural technician – someone who looks at the architect's information and makes drawings that can be used by the builder
- building technician – someone who is responsible for estimating the cost of the work and materials and general site management
- quantity surveyor – someone who calculates ongoing costs and payment for work done.

Building craft workers

Building craft workers are skilled people who work with materials to physically construct the building. The following jobs fall into this category:

- carpenter or joiner – someone who works with wood but also other construction materials such as plastic and iron. A carpenter primarily works on site while a joiner usually works off site, producing components such as windows, stairs, doors, kitchens, and **trusses**, which the carpenter then fits into the building
- bricklayer – someone who works with bricks, blocks and cement to build the structure of the building
- plasterer – someone who adds finish to the internal walls and ceilings by applying a **plaster skim**. They also make and fix plaster **covings** and plaster decorations

Key terms

Trusses – prefabricated components of a roof which spread the load of a roof over the outer walls and form its shape

Plaster skim – a thin layer of plaster that is put on to walls to give a smooth and even finish

Coving – a decorative moulding that is fitted to the top of a wall where it meets the ceiling

- painter and decorator – someone who uses paint and paper to decorate the internal plaster and timberwork such as walls, ceilings, windows and doors, as well as **architraves** and **skirting**

- electrician – someone who fits all electrical systems and fittings within a building, including power supplies, lights and power sockets

- plumber – someone who fits all water services within a building, including sinks, boilers, water tanks, radiators, toilets and baths. The plumber also deals with lead work and rainwater fittings such as guttering

- slater and tiler – someone who fits tiles on to the roof of a building, ensuring that the building is watertight

- woodworking machinist – someone who works in a machine shop, converting timber into joinery components such as window sections, spindles for stairs, architraves and skirting boards, amongst other things. They use a variety of machines such as lathes, bench saws, planers and sanders.

Building operatives

There are two different building operatives working on a construction site.

- specialist building operative – someone who carries out specialist operations such as dry wall lining, asphalting, scaffolding, floor and wall tiling and glazing

- general building operative – someone who carries out non-specialist operations such as kerb laying, concreting, path laying and drainage. These operatives also support other craft workers and do general labouring. They use a variety of hand tools and power tools as well as **plant** such as dumper trucks and JCBs.

> **Key terms**
>
> **Architrave** – a decorative moulding, usually made from timber, that is fitted around door and window frames to hide the gap between the frame and the wall
>
> **Skirting** – a decorative moulding that is fitted at the bottom of a wall to hide the gap between the wall and the floor

> **Key term**
>
> **Plant** – industrial machinery

The building team

Constructing a building or structure is a huge task that needs to be done by a team of people who all need to work together towards the same goal. The team of people is often known as the building team and is made up of the following people:

Client

The client is the person who requires the building or refurbishment. This person is the most important person in the building team because they finance the project and without the client there is no work. The client can be a single person or a large organisation.

Architect

The architect works closely with the client, interpreting their requirements to produce contract documents that enable the client's wishes to be realised.

Clerk of works

Selected by the architect or client to oversee the actual building process, the clerk of works ensures that construction sticks to agreed deadlines. They also monitor the quality of workmanship.

Local Authority

The Local Authority is responsible for ensuring that construction projects meet relevant planning and building legislation. Planning and building control officers approve and inspect building work.

Quantity surveyor

The quantity surveyor works closely with the architect and client, acting as an accountant for the job. They are responsible for the ongoing evaluation of cost and interim payments from the client, establishing whether or not the contract is on budget. The quantity surveyor will prepare and sign off final accounts when the contract is complete.

Specialist engineers

Specialist engineers assist the architect in specialist areas such as civil engineering, structural engineering and service engineering.

Health and safety inspectors

Employed by the Health and Safety Executive (HSE), health and safety inspectors ensure that the building contractor fully implements and complies with government health and safety legislation. For more information on health and safety in the construction industry, see Unit 1001 (page 2–10).

Building contractors

The building contractors agree to carry out building work for the client. Contractors will employ the required workforce based on the size of the contract.

Estimator

The estimator works with the contractor on the cost of carrying out the building contract, listing each item in the bill of quantities (e.g. materials, labour and plant). They calculate the overall cost for the contractor to complete the contract, including further costs as overheads such as site offices, management administration and pay, not forgetting profit.

Site agent

The site agent works for the building contractor and is responsible for the day-to-day running of the site such as organising deliveries.

Suppliers

The suppliers work with the contractor and estimator to arrange the materials that are needed on site and ensure that they are delivered on time and in good condition.

General foreman

The general foreman works for the site manager and is responsible for coordinating the work of the ganger (see below), craft foreman and subcontractors. They may also be responsible for the hiring and firing of site operatives. The general foreman also liaises with the clerk of works.

Craft foreman

The craft foreman works for the general foreman organising and supervising the work of particular crafts. For example, the carpentry craft foreman will be responsible for all carpenters on site.

Ganger

The ganger supervises general building operatives.

Chargehand

The chargehand is normally employed only on large building projects, being responsible for various craftsmen and working with joiners, bricklayers, and plasterers.

Operatives

Operatives are the workers who carry out the building work, and are divided into three subsections:

1. Craft operatives are skilled tradesman such as joiners, plasterers and bricklayers.
2. Building operatives include general building operatives who are responsible for drain laying, mixing concrete, unloading materials and keeping the site clean.
3. Specialist operatives include tilers, pavers, glaziers, scaffolders and plant operators.

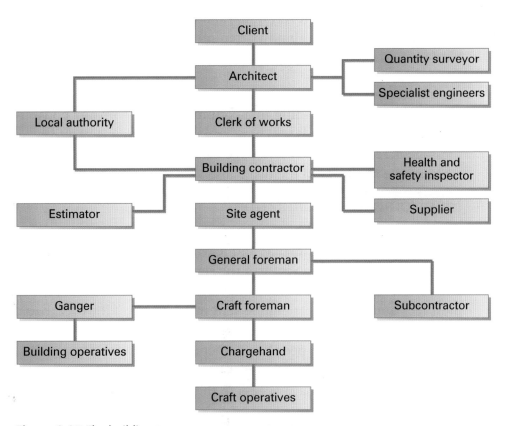

Figure 2.27 The building team

Methods of communication

There are many different ways of communicating with others and they all generally fit into one of these four categories:

1. speaking (verbal communication), for example talking face to face or over the telephone
2. writing, for example sending a letter or taking a message
3. body language, for example the way we stand or our facial expressions
4. electronic, for example email, fax and text messages.

Verbal communication

Verbal communication is instant, easy and can be repeated or rephrased if the message is not understood. However, it can be easily forgotten, or changed if it is passed on to other people. Accent and slang can make it harder to understand.

Written communication

This provides physical evidence of the message and can be passed on without being changed. It can also be read again if it is not understood. However, it takes longer to arrive and can be misunderstood or lost. Handwritten messages can sometimes be hard to read.

Body language

This can add extra meaning to a communication – if your body language is negative, this will have an impact on how a positive message is received. Body language can be quick and effective – for example waving and hand signals.

Electronic communication

This takes part the best parts of verbal and written communication, being instant and leaving a record of the message. E-mail can also inform the sender if the message has been sent and received. E-mail attachments also mean larger messages and documents can be sent immediately. However, bad signal or flat battery on a mobile phone or wi-fi can stop messages from being sent. Some people are uncomfortable with electronic messages, and e-mail needs the person receiving the message to be at a computer and have access to their account.

Other types of communication

Meetings and performance reviews

Meetings are used to pass on information face to face. Meetings can be very useful as they allow information on things such as safety or job progress to be passed both ways – from the person leading the meeting and those attending the meeting. The main downside to meetings is that for the duration of the meeting no actual work is getting done.

Performance reviews are done usually as a one to one and allows a line manager to inform the worker of how they are performing in their job. The performance review will record both areas where the worker has done well and areas where they may need to improve. The review should also suggest methods for informing – for example the worker may need more training or might want to have opportunities to progress their career.

The good thing about reviews is that they can be used to stop future problems from occurring by discussing any failings that the worker may have. They also allow the manager to praise the individual which may have a positive effect. The main negative side of reviews is that the worker may react badly to criticism which may affect their work in the future.

Media and posters/signs

Using media such as posters or signs is a good way of communicating specific information without the need for face-to-face or verbal communication. Safety signs are a good example of this, and a correctly displayed safety sign will inform all who pass it of a particular issue. The downside to media is that not everyone will see the information if it is poorly positioned or poorly designed. A poster or sign that is too cluttered or unclear will often just cause confusion.

Which type of communication should I use?

Of the many different types of communication, the type you should use will depend on the situation. If someone needs to be told something formally, then written communication is generally the best way. If the message is informal, verbal communication is usually acceptable.

The way that you communicate will also be affected by the person you are communicating with. You should of course always communicate in a polite and respectful manner with anyone you have contact with, but you must also be aware of the need to sometimes alter the style of your communication. For example, when talking to a friend, it may be fine to talk in a very informal way and use slang, but in a work situation with a client or a colleague, it is best to alter your communication to a more formal

Did you know?

Meetings can be an informal 30 minute gathering over a cup of tea or last a full day

style in order to show professionalism. In the same way, it may be fine to leave a message or send a text to a friend that says 'C U @ 8 4 wk', but if you wrote this for a work colleague or a client, it would not look very professional and they may not understand it.

Communication and teamwork

When working in a team communication is extremely important as you must be aware of what other members of the team are doing. This is because what you do will not only affect your working, but theirs as well, and vice versa. Even simple things like lifting in a group can lead to injuries if there is no communication between the people involved – for example, one person may begin lifting when the others are not ready.

Communicating what has been done or needs to be done is also important. Duplication of labour is a big problem in teams that do not communicate well – you don't want to get all the materials and tools ready for a task only to discover when you get there that another member of the team has already done it. Effective communication can prevent this from happening and can also help your team to work together more efficiently.

Communicating with other trades

Figure 2.28 You will work with people from other trades

Communicating with other trades is vital because they need to know what you are doing and when, and you need to know the same information from them. Poor communication can lead to delays and mistakes, and both can be costly. It is all too easy for poor communication to lead to work having to be stopped or redone. Imagine you are decorating a room in a new building. You are just about to finish when you find out that the electrician, plumber and carpenter all have work to finish in the room. This information didn't reach you and now the decorating will have to be redone once the other work has been finished. What a waste of time and money. A situation like this can be avoided with good communication between the trades.

Common methods of communicating in the construction industry

A career in construction means that you will often have to use written documents such as drawings, specifications and schedules. These documents can be very large and seem very complicated but, if you understand what they are used for and how they work, using them will soon become second nature.

The reasons for clear communication

Clear communication is vital for efficient relations between everyone who may be involved in a business, from the employer and employees through to clients and suppliers. You will have seen throughout this unit that there are several different methods to ensure good communication, and that there are procedures in place to avoid major problems developing in the workplace due to unclear communication.

Most of the crucial moments when you will need to use good, clear and effective communication relate to decisions that will have a wider effect on the business and those working around you. Some examples of these include:

- **alterations to drawings** – it is important to communicate any changes to these to everyone involved, as all the planning, estimating, material orders and work programmes will be based in part on these drawings. Not communicating changes could lead to mistakes in all these areas
- **variations to contracts** – the contract with the client is the crucial document that dictates all decisions that are made on a worksite. Changes to this document must be made known throughout a business
- **risk assessments** – the results of these assessments have a direct impact on the safety of workers on site, and should be made known to all
- **work restrictions** – these should be communicated to everyone as a restriction is put in place for a specific reason. The restrictions may be put in place for safety reasons. This would mean the area is unsafe so everyone who may be affected needs to be told.

Functional skills

Planning drawings for construction will give you the opportunity to practise the interpreting elements of functional skills, e.g. FM 1.3.1: Judge whether findings answer the original problem; FM 1.3.2: Communicate solutions to answer practical problems.

Working life

James is about to draw a kitchen plan for a client. Where would you begin if you were in his shoes?

Think about who to consult when planning a drawing. When talking to the client, what should James ask about? Appliances, the positioning of electrical points, and the way the client intends the kitchen to be used are as important as budget and design choices.

What other considerations are there? James will need to take into account openings for doors and windows, as well as supplies for essential services like water, gas, and electricity.

What other things would you include when drawing up plans?

FAQ

How many different forms are there?

Many forms are used in the building industry and some companies use more than others. You should ensure you get the relevant training on completing any form before using it.

How do I know what scale the drawing is at?

The scale should be written on the title panel (the box included on a plan or drawing giving basic information such as who drew it, how to contact them, the date and the scale).

How do I know if I need a schedule?

Schedules are only really used in large jobs where there is a lot of repeated design information. If your job has a lot of doors, windows etc., it is a good idea to use one.

Check it out

1   What information must be in your contract of employment? Write a list, explaining what each piece of information means.
2   Explain who draws up the plan, and their role in the construction process.
3   State three different types of drawings and give a suitable scale for each one. Draw up an example of each type of drawing you have selected.
4   State three of the main contract documents, and explain briefly what is important about each of them and the information they contain.
5   Explain the main purpose of a specification, and the information it contains.
6   What is the purpose of the bill of quantities?
7   What is a penalty clause? Describe a situation where a penalty clause might be used.
8   What is a confirmation notice?
9   Why, when calculating quantities of materials, do you allow for wastage? Give an example of wastage on site and calculate how you might need to take this into account.
10 Name a job in each of the four construction employment areas: professional; technician; building craft worker; building operative. Prepare a brief job description for the job you have selected.
11 Explain why the client is the most important member of the building team.
12 Show with the aid of a sketch how the area and circumference of a circle is worked out.

Getting ready for assessment

The information contained in this unit, as well as continued practical assignments that you will carry out in your college or training centre, will help you with preparing for both your end of unit test and the diploma multiple-choice test. It will also aid you in preparing for the work that is required for the synoptic practical assignments.

Working with contract documents such as drawings, specifications and schedules is something that you will be required to do within your apprenticeship and even more so after you have qualified.

You will need to be familiar with:

- interpreting building information
- determining quantities of material
- relaying information in the workplace.

To get all the information you need out of these documents you will need to build on the maths and arithmetic skills that you learnt at school. These skills will give you the understanding and knowledge you will need to complete many of the practical assignments, which will require you to carry out calculations and measurements.

You will also need to use your English and reading skills. These skills will be particularly important, as you will need to make sure that you are following all the details of any instructions you receive. This will be the same for the instructions you receive for the synoptic test, as it will for any specifications you might use in your professional life.

Communication skills have been a particular focus of this unit, and of learning outcome three. This unit has shown who the key personnel in the communication cycle are and demonstrated how poor communication between these people can have a negative effect on business and teamwork. You will need to be sure that you follow these guidelines for clear communication. Teamwork is a very important part of all construction work and can help work to run smoothly and ensure people's safety.

This unit has also explained the advantages and disadvantages of different types of communication. You will need to make sure that you always choose the most appropriate method of communication for the situation you are in. You also need to be confident in using all the different methods (letters, e-mail, telephone, signs, etc.) of communication.

You have seen that clear communication is vital for teams to work effectively. You will need to be able to demonstrate how the key personnel should communicate effectively.

The communicational skills that are explained within the unit are also vital in all tasks that you will undertake throughout your training and in life.

Good luck!

CHECK YOUR KNOWLEDGE

1 The abbreviation *ct* stands for:
   a cast iron.
   b cement.
   c column.
   d concrete.

2 Which of the following abbreviations is a timber?
   a pbd.
   b swd.
   c pvc.
   d fnd.

3 The purpose of a specification is:
   a to tell you how long a job will take.
   b to tell you the quality of work, and sizes not shown on the drawing.
   c to tell you the quantity of materials you will need.
   d to tell you how much the job will cost.

4 How is the area of a circle calculated?
   a πd^2
   b $d\pi^2$
   c πr^2
   d $r\pi^2$

5 A written warning is a:
   a mediation.
   b disciplinary.
   c procedure.
   d hierarchy.

6 Day work rate means that you get paid:
   a at the end of every day.
   b the same amount no matter what you do.
   c only for the work you do.
   d weekly.

7 What site paperwork is used to record hours worked?
   a day work sheet.
   b job sheet.
   c timesheet.
   d variation order.

8 To what scale are site plans usually drawn?
   a 1:500 or 1:200.
   b 1:2500 or 1:1250.
   c 1:100 or 1:50.
   d 1:10 or 1:5.

9 The abbreviation 'pbd' as used on a drawing represents:
   a plasterboard.
   b plaster.
   c polythene.
   d polyvinyl acetate.

10 What is the contract document that deals with repeated design information?
   a specification.
   b schedule.
   c plans and drawings.
   d conditions of contract.

UNIT 2003

Knowledge of building methods and construction technology 2

Whatever type of building is being constructed, there are certain principles and elements that must be included. For example, both a block of flats and a warehouse have a roof, walls and a floor.

These basic principles are applied across all the work carried out in construction and will apply to nearly all the projects you could work on. The primary areas this unit will look at are the principles behind walls, floors, roofs and internal work. This unit also supports NVQ Unit VR02 Conform to Efficient Work Practices.

This unit contains material that supports TAP Unit 2: Set Out for Masonry Structures. It also contains material that supports the delivery of the five generic units.

This unit will cover the following learning outcomes:

- Principles behind walls, floors and roofs
- Principles behind internal work
- Material storage and delivery of building materials.

Key term

Stress – a constant force or system of forces exerted upon a body resulting in strain or deformation

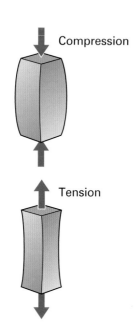

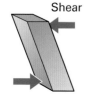

Figure 3.1 The three types of stress

K1. Principles behind walls, floors and roofs

All construction work requires working plans and drawings to complete the work. Drawings were covered earlier in Unit 2002 (pages 52-57).

It is important to realise that when buildings are designed at the planning stage, the different types of materials, methods of construction and principles behind all components are discussed and the most suitable materials and methods are chosen.

Principles of construction

Before looking at the range of different structures it is important to understand some of the key principles of construction. These are insulation and structural stability.

Structural loading

The main parts of a building that are in place to carry a load are said to be in a constant state of **stress.**

There are three main types of stress:

- Tension pulls or stretches a material and can have a lengthening effect.
- Compression squeezes the material and can have a shortening effect.
- Shear occurs when one part of a component slips or slides over another causing a slicing effect.

To cause one of these types of stress a component or member must be under the strain of a load. Within construction there are two main types of loading:

- dead load – the weight of the building itself and the materials used to construct the building, including components such as floors and roofs
- imposed loads – any moveable loads such as furniture, as well as natural forces such as wind, rain and snow.

To cope with the loads that a building must withstand there are load-bearing structural members strategically placed throughout the building.

There are three main types of load bearing members:

- horizontal members – one of the most common types of horizontal member is a floor joist, which carries the load and transfers it back to its point of support. The horizontal member, when under loading, can bend and be in all three types of stress, with the top in compression, the bottom in tension and the ends in shear. The bending can be contained by using correctly stress-graded materials or by adding a load-bearing wall to support the floor.

- vertical members – any walls or columns that are in place to transfer the loads from above (including from horizontal members) down to the substructure and foundations have vertical members. Vertical members are usually in a compression state

- bracing members – bracing members are usually fitted diagonally to form a triangle which stiffens the structure. Bracing members can be found in roofs and even on scaffolding. Bracing is usually in compression or tension.

Damp proof course (DPC)

With sustainability and energy efficiency being talked about more and more, the need to ensure that construction work is done with this in mind is vital. One of the main ways of ensuring that energy efficiency is maintained is by using correct insulation and a damp proof course.

A damp proof course (often shortened to DPC) or damp proof membrane (DPM) is a layer of non-absorbent material bedded on to a wall to prevent moisture penetrating into a building. There are three main ways moisture can penetrate into a building:

1. rising up from the ground
2. through the walls
3. moisture running downwards from the top of walls around openings or chimneys.

There are three types of DPC:

1. flexible
2. semi-rigid
3. rigid.

Flexible DPC

Flexible DPC comes in rolls of various widths to suit requirements. Nowadays most rolls are made of pitch-polymer or polythene but bitumen can still be found. Metal can be used as a DPC (in copper and lead) but because of the cost is mainly used in specialised areas. The most widely used and economic DPC material is polythene. Flexible DPC should always be laid upon

> **Remember**
>
> Where you are in the country will determine what materials you use for constructing. For example, some places with a lot of snowfall will require stronger structures to deal with the extra load from the snow

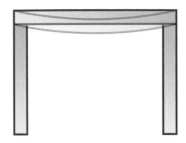

Figure 3.2 Horizontal structural members

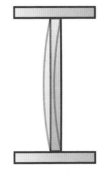

Figure 3.3 Vertical structural members

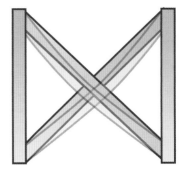

Figure 3.4 Bracing structural members

a thin bed of mortar and lapped by a minimum of 100 mm on a corner or if joining a new roll.

Semi-rigid DPC

This type of DPC is normally made from blocks of asphalt melted and spread in coats to form a continuous membrane for tanking for basements or underground work.

Rigid DPC

Rigid DPC uses solid material such as engineering bricks or slate, which were the traditional materials used. Slate is more expensive to use than other DPC materials and has no flexibility. If movement occurs the slate will crack, allowing damp to penetrate. Engineering bricks could be used for a garden wall if a DPC was required.

Modern insulation materials

Insulation, or more correctly thermal insulation, is a general term used to describe products that reduce heat loss or heat gain by providing a barrier between areas that are significantly different in temperature. There are a number of items in the home that benefit from insulation such as central heating boilers and hot water pipes. However buildings themselves need some extra help to make them more energy efficient. Home insulation therefore reduces the amount of heat that escapes from a building in the winter and protects it from getting too warm in the summer.

There are several different types of insulation available. These include:

- **rigid foam** – this has a high compressive strength and is usually used where it needs to support weight (e.g. under a floor, or in lofts as a storage solution). As well as strength and durability, rigid foam insulation can also provide additional properties such as fire resistance and acoustic insulation to minimise the level of sound travelling through walls and floors
- **sheep's wool** – this is a fairly new insulation product. However, before it is used as an insulation material, it must undergo an intensive cleaning process in order to remove the dirt and oils from the wool. The chemicals and energy used during this process must be taken into account when measuring its environmental impact, putting sheep's wool behind mineral wool in terms of its overall eco-credentials
- **polystyrene boards** – a high-density polystyrene sheet is usually cut to fit between ceiling joists. Safe and easy to install,

Unit 2003 Knowledge of building methods and construction technology 2 these boards easily meet the required insulation values with minimal board thickness.

Mineral wool insulation is the most common type of insulation.

Mineral wool insulation

There are two main types of mineral wool insulation:

- **glass mineral wool** – the world's most popular and widely used insulation material, glasswool is made from recycled glass bottles. This means it is ultra-eco-friendly. It is easy to handle and install, plus it is the most cost-effective insulation available
- **rock mineral wool** – rock mineral wool has a more solid structure so it is ideal for situations where it may be under compression, e.g. on a flat roof.

Mineral wool insulation products are available in rolls of different widths and thicknesses for quick and simple DIY installation, for example between the rafters in a roof or joists in a ceiling or floor.

It can be produced as lightweight 'slabs' for installing into the cavity walls when building new houses. 'Loose' mineral wool can also be used to fill cavity walls and is blown in through a hole drilled in the wall after it is built.

Installing mineral wool insulation is also an effective fire safety measure, as it does not burn easily, and can therefore prevent fire spreading. In fact, rock mineral wool can resist temperatures above 1000°C.

Other methods of insulation

As well as the insulation detailed above, certain other areas also require insulation. This can range from installing special double glazing units with thermal properties to ensuring that water pipes are correctly insulated (**lagged**).

Insulating water pipes is vital as it prevents them from freezing. Freezing can cause the pipes to crack and, in worst case scenarios, flood an entire building. Pipes are lagged either with a foam tube that slips over the pipe or with a mineral/rock wool blanket that is wrapped around them.

Cavity wall insulation is mentioned above in the cavity wall section but older buildings may not have insulation in the cavity. In these instances retro fitting of insulation is done by drilling holes into the exterior leaf of the cavity, filling it with a loose fill mineral wool and then resealing the holes.

Figure 3.5 Glass mineral wool

Figure 3.6 Rock mineral wool

Did you know?

There are in place at the moment various government grants in conjunction with energy providers that allow a house to be insulated for a minimal charge or even for free

This is current but for how long?

Substructure

All buildings start with the substructure, i.e. all the structure below ground level, up to and including the DPC. The substructure receives the loads from the main building and transfers them safely down to a suitable load-bearing layer of ground.

Setting out

Setting out refers to the marking out and positioning of a building. It is a very important operation as the setting out of a building must be as accurate as possible. Mistakes made at the setting out stage can prove very costly later. To appreciate the need for careful and accurate setting out, we have to understand and visualise the finished building and its requirements. When setting out a building, you must make sure that it is in the right place, it is level and it is square.

Planning foundations

Before any foundation can be formed, there must be extensive exploration, testing and preparation of the ground on which the building is to be erected.

As an apprentice, it is unlikely that you will find yourself involved in the design process of foundations for a particular building or structure, as this will be left to the architect and the structural design team. However, an appreciation of factors influencing the design of any foundation will help you to understand why there are different foundation types.

As already stated, the design of any foundation depends on a number of important factors:

- ground conditions
- soil type
- location of drains in relation to the proposed structure
- location of trees in relation to the proposed structure
- combined loads to be put on the ground directly beneath the proposed structure.

The purpose of foundations

The foundations of a building ensure that all **dead** and **imposed loads** are safely absorbed and transmitted through to the natural foundation or sub-soils on which the building is constructed. Failure to adequately absorb and transmit these loads will result in the stability of the building being compromised, and will undoubtedly cause structural damage.

Did you know?

The natural foundation is also referred to as the sub-foundation

Find out

Do some research into the types of load that may be imposed on a building structure

Key terms

Dead load – the weight of the structure

Imposed load – the additional weight/loading that may be placed upon the structure

Foundations must also be able to allow for ground movement brought about by shrinkage or expansion of the soil as it dries out or becomes wet. The severity of shrinkage or expansion depends on the type of soil being built on.

Frost may also affect ground movement, particularly in soils that hold water for long periods. Freezing of this retained water can cause expansion of the sub-soil.

Find out

What is the term given to the effect on sub-soil when retained water freezes?

Types of soil

As you can imagine there are many different types of soil. For foundation design purposes, these have been categorised as follows:

- rock
- gravel
- clay
- sand
- silt.

Each of these categories of sub-soil can be broken down even further, for example,

- clay which is sandy and very soft in its composition
- clay which is sandy but very stiff in its composition.

This information will be of most interest to the architect, but nonetheless is of the utmost importance when designing the foundation of a building.

A number of calculations are used to determine the size and make-up of the foundation. These calculations take into account the **bearing capacity of the soil**. Calculations for some of the more common types of foundations can be found in the current *Building Regulations*. However, these published calculations cannot possibly cover all situations. Ultimately it will be down to the expertise of the building design teams to accurately calculate the bearing capacity of the soil and the make-up of the foundation.

Key term

Bearing capacity of soil – the load that can be safely carried by the soil without any adverse settlement

In the early stages of the design process, before any construction work begins, a site investigation will be carried out to ascertain any conditions, situations or surrounding sites which may affect the proposed construction work. A great deal of data will need to be gathered during site investigations, including:

- position of boundary fences and hedges
- location and depth of services, including gas, electricity, water, telephone cables, drains and sewers

Find out

Look at the different methods and equipment used to locate and identify various hidden services

Find out

How are the different soil tests carried out?

Did you know?

Site investigations or surveys will also establish the contours of the site. This will identify where certain areas of the site will need to be reduced or increased in height. An area of the site may need to be built up in order to mask surrounding features outside the boundaries of the proposed building project

Find out

How can plant growth affect some structures?

- existing buildings which need to be demolished or protected
- position, height, girth and spread of trees
- types of soil and the depths of these various soils.

The local authorities will normally provide information relating to the location of services, existing buildings, planning restrictions, preservation orders and boundary demarcation. However, all of these will still need to be identified and confirmed through the site investigation. In particular, hidden services will need to be located with the use of modern electronic surveying equipment.

Soil investigations are critical. Samples of the soil are taken from various points around the site and tested for their composition and for any contamination. Some soils contain chemicals that can seriously damage the foundation concrete. These chemicals include sodium and magnesium sulphates. The effects of these chemicals on the concrete can be counteracted with the use of sulphate-resistant cements.

Many different tests can be carried out on soil. Some are carried out on site; others need to be carried out in laboratories. Tests on soil include:

- penetration tests – to establish density of soil
- compression tests – to establish shear strength of the soil or its bearing capacity
- various laboratory tests – to establish particle size, moisture content, humus content and chemical content.

Once all site investigations have been completed and all necessary information and data has been established in relation to the proposed building project, site clearance can take place.

Site clearance

The main purpose of site clearance is to remove existing buildings, waste, vegetation and, most importantly, the surface layer of soil referred to as topsoil. It is necessary to remove this layer of soil, as it is unsuitable to build on. This surface layer of soil is difficult to compact down due to the high content of vegetable matter, which makes the soil soft and loose. The topsoil also contains various chemicals that encourage plant growth, which may adversely affect some structures over time.

The process of removing the topsoil can be very costly in terms of both labour and transportation. The site investigations will determine the volume of topsoil that needs to be removed.

In some instances, the excavated topsoil may not be transported off site. Where building projects include garden plots, the topsoil may need to be stored on site, thus reducing excessive labour and transportation costs. However, where this is the case, the topsoil must be stored well away from areas where buildings are to be erected or materials are to be stored, to prevent contamination of soils or materials.

Once the site clearance is complete, excavations for the foundations can start and the concrete foundations can be constructed.

Construction of concrete foundations

Trench excavation

In most modern-day construction projects, trenches are excavated by mechanical means. Although this is an expensive method, it reduces labour time and the risks associated with manual excavation work. Even with the use of machines to carry out excavations, an element of manual labour will still be needed to clean up the excavation work: loose soil from both the base and sides of the trench will have to be removed, and the sides of the trench will have to be finished vertically.

Manual labour is still required for excavating trenches on some projects where machine access is limited and where only small strip foundations of minimum depths are required.

Trenches to be excavated are identified by lines attached to and stretched between profiles. This is the most accurate method of ensuring trenches are dug to the exact widths required.

Trench support

The type and extent of support required in an excavated trench will depend predominantly on the depth of the trench and the stability of the soil.

Traditionally, trench support was provided by using varying lengths and sizes of timber, which can easily be cut to required lengths. However, timber can become unreliable under certain loadings, pressures and weather conditions and can fail in its purpose.

More modern types of materials have been introduced as less costly and time-consuming methods of providing the required support. These materials include steel sheeting, rails and props. Trench support can be provided with a mixture of timber and steel components.

Here you can see the methods of providing support in trenches with differing materials and a combination of these materials.

Figure 3.7 Removing soil from a site

Did you know?

The Health and Safety Executive (HSE) has produced detailed documents that deal exclusively with safety in excavations. These documents can be downloaded from the HSE website or obtained directly from the HSE upon request.

Regulations relating to safety in excavations are set out in the Construction Regulations and these must be strictly adhered to during the work

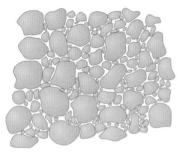

Well graded

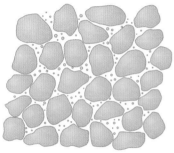

Poorly graded

Figure 3.8 Well graded and poorly graded concrete

Concrete

Concrete is a mixture of cement, aggregates and water. The aggregate is normally free and coarse.

The coarse aggregate is the bulk of the concrete, while the fine aggregate fills in the voids between the larger particles.

The cement is the binder that holds all the aggregates together. Water is required to cause a chemical reaction (**hydration**) that changes the dry cement powder into an adhesive.

Fine aggregates

Fine aggregates can be obtained from riverbeds or sandpits, or dredged from the seabed. Dredged aggregates must be washed to remove any mud and weed. Sea sand must not be used due to the high salt content.

The shape and size of the grains can affect the strength of the concrete. The particles should be irregular in shape (not rounded) and well graded, with the size not larger than 5 mm. Also, if the grain size is too small it will increase the total surface area so that the designed quantity of cement becomes insufficient, causing a weakness in the finished mix.

Clay and silt particles prevent the cement from bonding to the aggregate, so each load delivered to the site should be tested for 'cleanliness'. The amount of silt must not be more than 10 per cent of the volume of aggregate.

Coarse aggregates

Coarse aggregates are the larger particles of a concrete mix. They can be made from gravel or crushed rock. If gravel is to be used, it should be crushed as the irregular shape of the crushed particles gives a better bond with the cement than if they are left smooth. The size of the coarse aggregate can range between 5 and 40 mm but if it is to be used for reinforced concrete, the size should be kept smaller than 20 mm.

Water

Water is used in the production of concrete to enable the cement to set and also to make the concrete 'workable'. Water must not contain any impurities, which might affect the strength of the concrete. The general rule for the quality of water is that it should be drinkable (potable).

Mixes

Concrete is designed for where it is to be used. Mass concrete for normal strip foundations should be mixed to a ratio of 1:3:6 of cement, fine aggregate and coarse aggregate. Concrete walls, beams and suspended floors, etc. should have a ratio of 1:2:4 of cement, fine aggregate and coarse aggregate.

Formwork

The purpose of formwork is to hold the freshly placed and compacted concrete until it has set. To achieve this, the formwork should:

- be rigid enough to prevent bending
- be strong enough to carry the weight of the concrete
- be set in place to line and level
- have tight joints to prevent water or cement paste loss
- have suitable size panels to allow safe and easy handling
- be designed in such a way that air pockets are not trapped.

Formwork may be made from timber or steel, and the choice of which to use usually depends on how many times the formwork is to be used. Steel formwork may be more expensive than timber initially but it can be reused more times as timber can be easily damaged during the striking of the formwork.

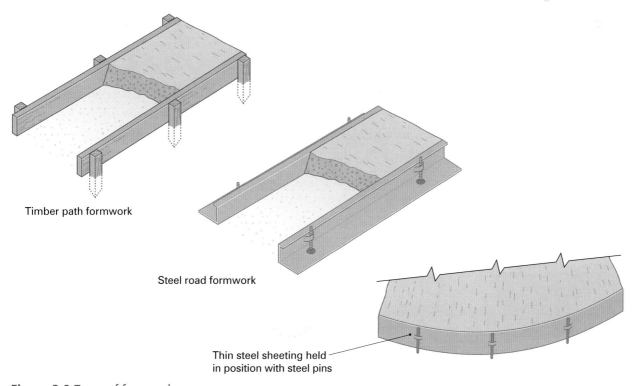

Timber path formwork

Steel road formwork

Thin steel sheeting held in position with steel pins

Figure 3.9 Types of formwork

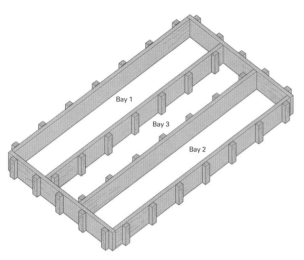

Figure 3.10 Alternative strip method used for large floor areas

Remember

Steel tools should not be used to clean formwork as they can easily scratch the forms, which could show on the concrete surface

Key terms

Compression – squeezing or squashing together

Tension – stretching

Figure 3.11 An example of concrete spacers

Formwork for ground floors

Floors for buildings such as factories and warehouses have large areas and would be difficult to lay in one slab. Floors of this type are usually laid in strips up to 4.5 m wide, running the full length of the building (see Figure 3.10). The actual formwork would be similar to that used for paths.

Fixing formwork

The fixing of formwork may be made using nails, clamps or bolts. Clamps or bolts are preferred over nails as they are easier to strip off the formwork after completion, with less chance of anyone stepping on a nail which has been left sticking out of an old piece of formwork.

Striking of formwork

Striking of formwork is the removal of the formwork from the hardened concrete. Vertical sides of the formwork may be removed after 12 hours while soffits supporting lintels, etc. should be left in place between 7 and 14 days.

All formwork should be cleaned as soon as it has been struck. A stiff brush should be used to remove any dust and cement grout. Stubborn bits of grout can be removed using a wooden scraper.

Reinforcement

Concrete is strong in **compression** but weak in **tension** so, to prevent concrete from being 'pulled' apart when under pressure, steel reinforcement is provided. The type and position of the reinforcement will be specified by the structural engineer.

The reinforcement must always have a suitable thickness of concrete cover to prevent the steel from rusting if exposed to moisture or air. The amount of cover required depends on the location of the site with respect to exposure conditions, and ranges from 20 mm in mild exposure to 60 mm for severe exposure to water.

To prevent the reinforcement from touching the formwork, spacers should be used. Made from concrete, fibre, cement or plastic they are available in several shapes and various sizes to give the correct cover.

Types of foundation

Concreting will be used to create a range of different foundation types. As previously stated, the design of a foundation will be down to the architect and structural design team. The final decision on the suitability and depth of the foundation, as well as the thickness of the concrete, will rest with the local authority's building control department.

Strip foundations

The most commonly used strip foundation is the 'narrow strip' foundation, which is used for small domestic dwellings and low-rise structures. Once the trench has been excavated, it is filled with concrete to within 4–5 courses of the ground level DPC. The level of the concrete fill can be reduced in height, but this makes it difficult for the bricklayer due to the confined area in which to lay bricks or blocks.

The depth of this type of foundation must be such that the soil acting as the natural foundation cannot be affected by the weather. This depth would not normally be less than 1 m.

The narrow strip foundation is not suitable for building with heavy structural loading or where the soil is weak in terms of supporting the combined loads imposed on it. Where this is the case, a wide strip foundation is needed.

Wide strip foundations

Wide strip foundations consist of steel reinforcement placed within the concrete base of the foundation. This removes the need to increase the depth considerably in order to spread heavier loads adequately.

Raft foundations

These types of foundation are used where the soil has poor bearing capacity, making the soil prone to settlement. A raft foundation consists of a slab of concrete covering the entire base of the structure. The depth of the concrete is greater around the edges of the raft in order to

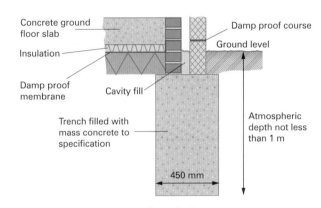

Figure 3.12 Narrow strip foundation

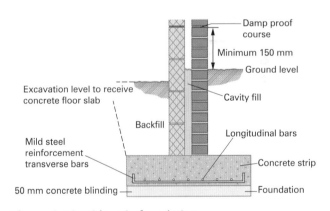

Figure 3.13 Wide strip foundation

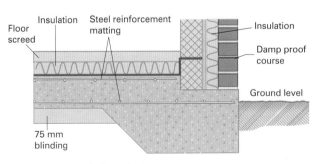

Figure 3.14 Raft foundation

protect the load-bearing soil directly beneath the raft from further effects of moisture taken in from the surrounding area.

Pad foundations

Pad foundations are used where the main loads of a structure are imposed only at certain points – for example where brick or steel columns support the weight of floors or roof members, and walls between the columns are of non-load bearing cladding panels. The simplest form of pad foundation uses individual concrete pads placed at various points around the base of the structure, with concrete ground beams spanning across and between them. The individual concrete pads will absorb the main imposed loads, while the beams will help to support the walls.

The depth of a pad foundation will depend on the load being imposed on it; in some instances, there may be a need to use steel reinforcement to prevent excessive depths of concrete. This type of pad foundation can reduce the amount of excavation work required, as trenches do not need to be dug out around the entire base of the proposed structure.

Piled foundations

There are many different types of piled foundation, each with an individual purpose in relation to the type of structure and ground conditions.

Short bored piled foundations are the most common piled foundations. They are predominantly used for domestic buildings where the soil is prone to movement, particularly at depths below 1 m.

A series of holes are bored by mechanical means around the perimeter of the base of the proposed building. The diameter of the bored holes will normally be between 250 and 350 mm and can extend to depths of up to 4 m. Once the holes have been bored, shuttering is constructed to form

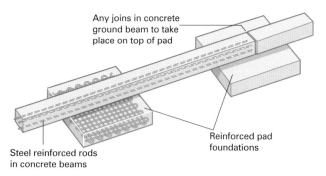

Any joins in concrete ground beam to take place on top of pad

Steel reinforced rods in concrete beams

Reinforced pad foundations

Figure 3.15 Square pad foundation with spanning ground beams

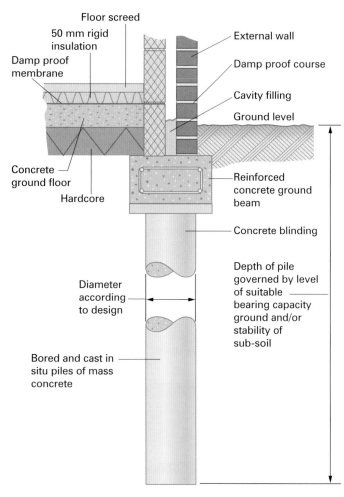

Floor screed

50 mm rigid insulation

Damp proof membrane

Concrete ground floor

Hardcore

Diameter according to design

Bored and cast in situ piles of mass concrete

External wall

Damp proof course

Cavity filling

Ground level

Reinforced concrete ground beam

Concrete blinding

Depth of pile governed by level of suitable bearing capacity ground and/or stability of sub-soil

Figure 3.16 Typical short bored piled foundation

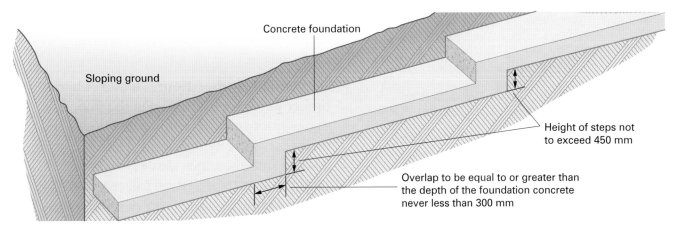

Figure 3.17 Typical stepped foundation

lightweight reinforced concrete beams, which span across the bored piles. The bored holes are then filled with concrete, with reinforcement projecting from the top of the pile concrete, so it can be incorporated into the concrete beams that span the piles.

As with pad foundation, short bored piled foundations can significantly reduce the amount of excavated soil because there is no need to excavate deep trenches around the perimeter of the proposed structure.

Stepped foundations

A stepped foundation is used on sloping ground. The height of each step should not be greater than the thickness of the concrete, and should not be greater than 450 mm. Where possible, the height of the step should coincide with brick course height in order to avoid oversized mortar bed joints and eliminate the need for split brick courses. The overlap of the concrete between any given step and the step below it should not be less than 300 mm or less than the thickness of the concrete.

Superstructure

The superstructure covers everything above the substructure, from walls to floors to roofing. The purpose of the superstructure is to enclose and divide space, as well as spread loads safely into the substructure.

Within the superstructure, you will find the primary, secondary and finishing elements, as well as the services.

Floors

There are two main types of floor: ground and upper.

Ground floors

There are a few main types of ground floor. These are the ones you will most often come across.

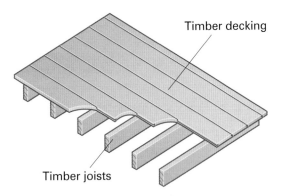

Figure 3.18 Suspended timber floor

Timber decking

Timber joists

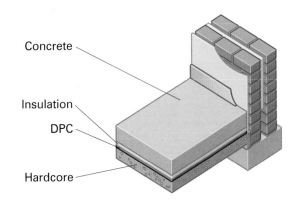

Figure 3.19 Solid concrete floor

Concrete

Insulation

DPC

Hardcore

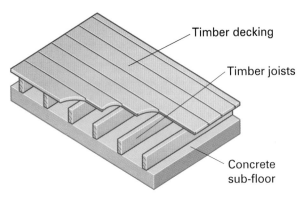

Figure 3.20 Floating floor

Timber decking

Timber joists

Concrete sub-floor

- **suspended timber floor** – a floor where timber joists are used to span the floor. The size of floor span determines the depth and thickness of the timber used. The joists may be built into the inner skin of brickwork, sat upon small walls (dwarf/sleeper wall), or suspended from some form of joist hanger. The joists should span the shortest distance and sometimes dwarf/sleeper walls are built in the middle of the span to give extra support or to go underneath load-bearing walls. The top of the floor is decked with a suitable material (usually chipboard or solid pine tongue and groove boards). As the floor is suspended, usually with crawl spaces underneath, it is vital to have air bricks fitted. These allow air to flow under the floor, preventing high moisture content and timber rot

- **solid concrete floor** – concrete floors are more durable and are constructed on a sub-base incorporating hardcore, damp proof membranes and insulation. The depth of the hardcore and concrete will depend on the building and will be set by the *Building Regulations* and the local authority. Underfloor heating can be incorporated into a solid concrete floor. Great care must be taken when finishing the floor to ensure it is even and level

- **floating floor** – basic timber floor construction that is laid on a solid concrete floor. The timbers are laid in a similar way to joists, though they are usually a maximum of 50 mm thick as there is no need for support. The timbers are laid on the floor at predetermined centres, and are not fixed to the concrete base (hence floating floor); the decking is then fixed onto the timbers. Insulation or underfloor heating can be placed between the timbers to enhance the thermal and acoustic properties.

Upper floors

Again, solid concrete slabs such as pre-cast beams can be used in larger buildings, but the most common type of upper floor is the suspended timber floor. As before, the joists are either built into the inner skin of brickwork or supported on some form of joist hanger. Spanning the shortest distance, with load-bearing walls acting as supports, it is vital that **regularised joists** are used, as a level floor and ceiling are required. The tops of the joists are again decked out, with the underside being clad in plasterboard and insulation placed between the joists to help with thermal and acoustic properties.

Walls

There are two main types of wall within a building: external and internal.

External walls

External walls come in a variety of styles, but the most common is cavity walling. Cavity walling is simply two brick walls built parallel to each other, with a gap – the cavity – between them. The cavity wall acts as a barrier to weather, with the outer leaf preventing rain and wind from penetrating the inner leaf. The cavity is usually filled with insulation to prevent heat loss.

Timber kit houses are becoming more and more common as they can be erected to a wind and watertight stage within a few days. The principle is similar to a cavity wall: the inner skin is a timber frame clad in timber sheet material and covered in a breathable membrane to prevent water and moisture penetrating the timber. The outer skin is usually face brickwork.

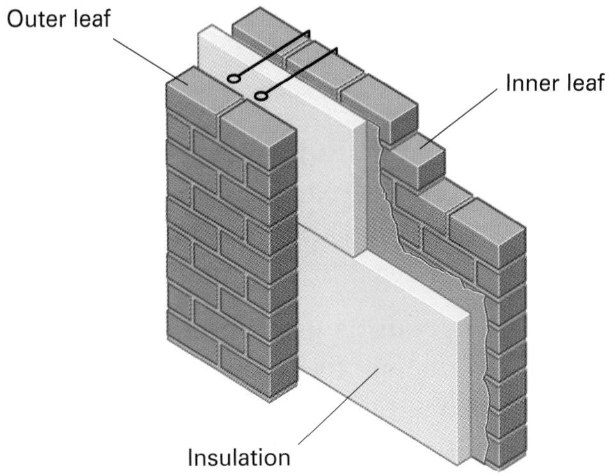

> **Did you know?**
>
> Sometimes metal beams are used to span floors and act as load-bearing members. These beams are either dipped in a chemical coating or painted to prevent rust

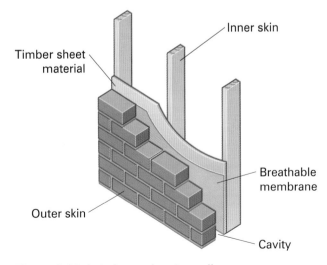

Figure 3.21 A cavity wall

Outer leaf

Inner leaf

Insulation

Figure 3.22 A timber and cavity wall

Inner skin

Timber sheet material

Breathable membrane

Outer skin

Cavity

Safety tip

Load-bearing walls must not be altered without first providing temporary supports to carry the load until the work has been completed

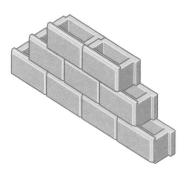

Figure 3.23 Solid block wall

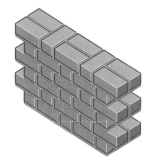

Figure 3.24 Solid brick wall

Figure 3.25 Timber stud wall

Figure 3.26 Metal stud wall

There are also other types of exterior walling such as solid stone or log cabin style. Industrial buildings may have steel walls clad in sheet metal.

Internal walls

Internal walls are either load-bearing or non-load bearing.

The difference between the two is that load-bearing walls will support the weight of any upper floors or roofs; in a house there will usually be at least one load-bearing wall. Non load-bearing walls are simply used to divide the area into separate rooms.

Internal walls come in a variety of styles. Here is a list of the most common types.

- **solid block walls** – simple blockwork walls, sometimes using fairface blockwork. These may be covered with plasterboard or plastered over to give a smooth finish, to which wallpaper or paint is applied. Solid block walls offer low thermal and sound insulation qualities; however, advances in technology mean that materials such as thermalite blocks can give better sound and heat insulation

- **solid brick walling** – usually made with face brickwork as a decorative finish. It is unusual for all walls within a house to be made from brickwork

- **timber stud walling** – more common in timber kit houses and newer buildings. Timber stud walling is also preferred when dividing an existing room, as it is quicker to erect. Clad in plasterboard and plastered to a smooth finish, timber stud partitions can be made more fire-resistant and sound/thermal qualities can be improved with the addition of insulation or different types of plasterboard. Another benefit of timber stud walling is that timber noggins can be placed within the stud to give additional fixings for components such as radiators or wall units. Timber stud walling can also be load-bearing, in which case thicker timbers are used

Figure 3.27 Ground lats

- **metal stud walling** – similar to timber stud, except that metal studs are used and the plasterboard is screwed to the studding

- **grounds lats** – timber battens that are fixed to a concrete or stone wall to provide a flat surface, to which plasterboard is attached and a plaster finish applied.

Cavity walling

Cavity walls are mainly used for house building and extension work to existing homes and flats. They consist of two separate walls built with a cavity between, joined together by metal ties. In most cases the outer wall is made of brick and the inner skin is made of block.

The main reason for this type of construction is to protect the inside from water penetration. The cavity forms a barrier – if the outer wall becomes wet (through rain, snow, etc.), water is not passed through because the two walls do not touch. Air circulating around the cavity dries the dampness caused, as well as keeping the inner wall dry. Where the walls do meet, for example at door and window openings, a damp proof course (DPC) is used to stop water penetration. The cavity can be insulated either partially or fully to make the building warmer and more energy-efficient.

Roofs

Although there are several different types of roofing, all roofs will technically be either a flat roof or a pitched roof.

Flat roofs

A flat roof is a roof with a pitch of 10 or less. The pitch is usually achieved through laying the joists at a pitch, or by using **firring pieces.**

The main construction method for a flat roof is similar to that for a suspended timber floor, with the edges of the joists being supported via a hanger or built into the brickwork, or even a combination of both. Once the joists are laid and firring pieces are fitted (if required), insulation and a vapour barrier are put in place. The roof is then decked on top and usually plasterboarded on the underside. The decking on a flat roof must be waterproof, and can be made from a wide variety of materials, including fibreglass or bitumen-covered boarding with felt layered on it.

Drainage of flat roofs is vital. The edge to which the fall leads to must have suitable guttering to allow rainwater to run away, rather than down the face of the wall.

> **Key terms**
>
> **Regularised joists** – joists that are all the same depth
>
> **Firring pieces** – tapered strips of timber

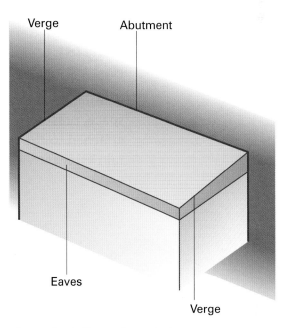

Figure 3.28 Flat roof terminology

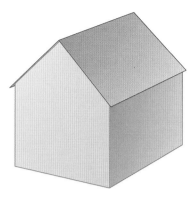

Figure 3.29 Duo pitch roof with gable ends

Did you know?

Due to the fact that heat rises, the majority of heat loss from a building is through its roof. Insulation such as mineral wool or polystyrene must be fitted to roof spaces and ideally to any intermediate floors as well

Pitched roofs

There are several types of pitched roof, from the basic gable roof to more complex roofs such as mansard roofs. Whichever type of roof is being fitted to a building, it will most likely be constructed in one of the following ways:

- **prefabricated truss roof** – as the name implies, this is a roof that has prefabricated members called trusses. Trusses are used to spread the load of the roof and to give it the required shape. Trusses are factory-made, delivered to site and lifted into place, usually by a crane. They are easy and quick to fit: either they are nailed to a wall plate or held in place by truss clips. Once fitted, bracing is attached to keep the trusses level and secure from wind. Felt is then fixed to the trusses and tiles or slate are used to keep the roof and dwelling waterproof

- **traditional/cut roof** – as an alternative to trusses, the cut roof uses loose timbers that are cut in-situ to give the roof its shape and to spread the relevant load. More time-consuming and difficult to fit than trusses, the cut roof uses rafters that are individually cut and fixed in place, with two rafters forming a sort of truss. Once the rafters are all fixed, the roof is finished with felt and tiles or slate.

Metal trusses can also be used for industrial or more complex buildings.

To finish a roof where it meets the exterior wall (eaves), you must fix a vertical timber board (fascia) and a horizontal board (soffit) to the foot of the rafters/trusses. The fascia and soffit are used to close off the roof space from insects and birds.

Ventilators are attached to the soffits to allow air into the roof space, preventing rot, and guttering is attached to the fascia board to channel rainwater into a drain.

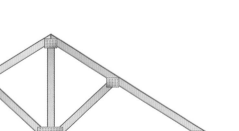

Figure 3.30 Prefabricated wooden roof truss

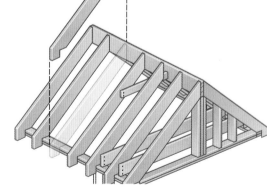

Figure 3.31 Individually cut rafters

Insulation materials in roofs

Due to the fact that heat rises, the insulation materials contained within a roof space, whether it be a flat or pitched roof, have a vital impact on the energy efficiency of the building.

Using the wrong type or size of insulation can vastly reduce the efficiency of the building. The type and size of insulation that should be used will be decided at the early planning stage and the architect should state what is to be used.

On smaller jobs where there is no architect, or on jobs where installing insulation is the only job, it is important that you read the manufacturer's information. This should be contained on the packaging, but if in doubt telephone the manufacturer to check.

Finishing elements

Finishing elements are the final surfaces of an element, which can be functional or decorative.

The main finishing elements are:

- plaster
- render
- paint
- wallpaper
- flooring
- felt and battens.

Finishing for walls

Plaster

Plaster can be used on a variety of wall surfaces to give a smooth and even finish. The plaster comes in powder form, usually bagged, and is mixed with water until it reaches a consistency that allows it to be applied to the surface and trowelled smooth. Ready-mix plaster is also available, but is more expensive, especially when many surfaces have to be plastered.

These are the main surfaces to which plaster is applied:

- brick/block work – prior to application, a bonding agent must be applied to the wall (usually a coat of watered down PVA), to help the plaster adhere to the surface. Usually a first coat of bonding plaster is applied to the wall to give it a level and flat surface; when this is dry, a second, finish coat is applied. As the finish coat is drying, the plasterer will work on the wall, smoothing it out until it is as smooth as glass

Safety tip

To prevent dermatitis, ensure that you wear gloves when working with plaster, render or cement

Key term

Key – the end result of a process that prepares a surface, usually by making it rough or grooved, so that paint or some other finish will stick to it

Figure 3.32 Lath and plaster

Figure 3.33 Plaster being applied and trowelled

- plasterboard – as plasterboard is a flat surface to begin with, a bonding coat is rarely used. Generally, the plasterboard is fixed with the back face (the face with the writing) exposed to give better adhesion. Whether it is a wall or a ceiling, the plasterer will again work the finish coat to a very smooth surface
- lath and plaster – this is usually found in old properties. The laths are thin strips of wood, which are fixed to the wall with small gaps between to give the plaster a **key.** Once the laths are fixed the plasterer will apply bonding and finish coats as before.

Plasterboard with a tapered edge can also be fixed to the walls. In this case, instead of plastering the entire wall, the plasterer will simply fill the nail/screw holes, fit tape where the plasterboard joins are, and fill only the joints. Pre-mixed plaster is usually used for this; when it is dry, a light sanding is required to give a smooth finish. This method is preferred in newer buildings, especially timber kit houses.

Not all walls are plastered smooth, as some clients may require a rough or patterned finish. Although not technically a plaster, Artex™ is often used to give decorative finishes, especially on ceilings.

Render

Render is similar to plaster in that it is trowelled onto brick or blockwork to give a finish. Applied to external walls, the render must be waterproof to prevent damage to the walls. Different finishes are available, from stippling to patterning.

Paint

Paint is applied to various surfaces and is available in many different types to suit specific jobs. Paint is applied for a variety of reasons, the most common being to:

- protect – steel can be prevented from corroding due to rust, and wood can be prevented from rotting due to moisture and insect attack
- decorate – the appearance of a surface can be improved or given a special effect (for example marbling and wood graining)
- sanitise – a surface can be made more hygienic with the application of a surface coating, preventing penetration and accumulation of germs and dirt, and allowing easier cleaning.

Paint is either water-based or solvent-based. When a paint is water-based, it means that the main liquid part of the paint is water; with a solvent-based paint, a chemical has been used instead of water to dissolve the other components of the paint.

Water-based paint is generally used on walls and ceilings, while solvent-based paint is used on timber mouldings, doors, metals, etc.

There are other surface finishes besides paint such as varnish (used on wood), masonry paint (used on exterior walls) and preservatives, which are used to protect wood from weather and insect attack.

Wallpaper

Wallpapers are used to decorate walls; thicker wallpapers can also hide minor defects. Basic wallpapers are made from either wood pulp or vinyl.

Wood-pulp papers can be used as preparatory papers or finish papers. Preparatory papers are usually painted with emulsion to provide a finish, or they can be used as a base underneath finish papers. Types of wood-pulp paper include plain, coloured and reinforced lining paper as well as wood chip.

Vinyl wallpaper is a hard-wearing wallpaper made from a PVC layer attached to a pulp backing paper. Types of vinyl paper include patterned, sculptured or blown vinyl.

Wallpaper is hung on a wall using a paste. Not all pastes have the same strength, so make sure you choose the correct paste for the type of paper you are using.

Finishing for roofs

Finishing for a roof requires the laying of felt over the trusses or rafters to provide a weatherproof barrier. On top of the felt, battens are placed horizontally along the roof. This not only helps to keep the felt in place but, more importantly, allows a fixing and gives an angle for either a tile or slate finish.

Figure 3.34 Batten and felt in a roof

Finishing for flooring

Flooring is traditionally finished with carpets in the majority of rooms, with a vinyl or linoleum material used in the kitchen and bathroom. Today, however, there is a larger choice in how floors are finished, with the main types of floor finish being:

- **carpet** – still preferred in a large number of homes, as it feels warmer and more comfortable than other finishes. Some carpets are more moisture resistant and can be fitted in any room
- **linoleum** – a plastic-based floor covering that is glued down and comes in a variety of patterns or colours. It is less popular nowadays, as many people prefer laminated flooring
- **laminated flooring** – a solid MDF style base with a thin layer,

or veneer, placed on top which can be wood grained or tiled. This is popular as it is hard-wearing, comes in short lengths which click together and is easily fitted. If fitting laminated flooring in an area of high moisture such as a kitchen or bathroom, ensure that that the flooring you buy is suitable

- **tile** – it has recently become popular, especially in kitchens, to use traditional large floor tiles. These are very hard-wearing and care needs to be taken when installing them to ensure that they are laid level.

Solid wood flooring

There are two methods in which a finish can be achieved for this type of flooring.

If you have old wooden tongue and groove floorboards, you can remove the existing floor covering, sand down the floorboards and cover them with a suitable varnish. This is not a very energy efficient method as heat will escape from the gaps between the floorboards.

Where chipboard flooring has been used, fitting solid tongue and groove boarding on top of the chipboard or instead of it will give you a more traditional look.

Figure 3.35 Chipboard flooring

K2. Principles behind internal work

Types of material

Materials were covered earlier on pages 103–105. You can refer back to this section for more information on the properties and storage of materials.

There are several types of material that you will use for internal work. These include the materials shown in the table below.

Material	Use
Polyurethane	A polymer that is used in a wide variety of building materials ranging from varnish to adhesive.
Glass fibre quilt	Used mainly to insulate areas within the buildings such as roof spaces and the areas between floors and walls.
Common brick	A basic brick that can be used in any type of walling.
Common block	A block made from concrete that can be used in any type of walling.
Aggregates	Covers a wide range of materials such as sand and chippings. Used in the creation of cements, concretes and renders.
Plasterboard	Man-made board that is used to clad walls and ceilings.

Concrete	Mixed from aggregates. Can be used to create formwork for floors or pillars, etc.
Metals	Used widely on large sites as a means of support; on some commercial/industrial buildings the entire skeleton of the frame can be made from metal.
Mineral wool	Similar uses to glass fibre quilt, acting as insulation.
Softwood	Used for a variety of joinery purposes from doors to stairs to roofs.
Hardwood	Used where a higher class of finish is needed (such as doors). Very rarely used in roofs unless the beams are exposed.
Facing brick	Used on the outer skin of a cavity wall; more decorative than common brick.
Thermal block	Used on the inner skin of a cavity wall to provide better thermal insulation.
Glass	Used in windows and doors to allow natural light to enter the premises.
Plaster	Used to cover plasterboard or brick/block walls to give a smooth finish ready for decoration.
Engineering brick	Used in the same ways as common or facing bricks but where more strength is required.

Internal components

Secondary elements are not essential to the building's strength or structure, but provide a particular function such as completing openings in walls.

The main secondary elements are:

- stairs
- frames and linings
- doors
- windows
- architrave and skirting.

Stairs

Stairs are used to provide access between different floors of a dwelling or to gain access to a higher/lower area. Stairs are made up of a number of steps, and each continuous set of steps running in the same direction is known as a flight. Steps are made of vertical boards called risers and horizontal boards called treads.

There are various types of stair, ranging from spiral staircases (often fitted where there is a lack of space) to multi-flight staircases such as dog-leg or half-turn stairs.

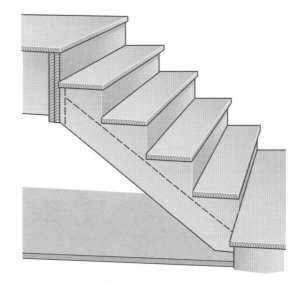

Figure 3.36 A simple staircase

Stairs are strictly governed by the *Building Regulations* and there are numerous requirements that must be adhered to when constructing and installing them.

Stairs are generally made from four types of material:

- timber – the most common material, used widely in almost all buildings
- in-situ-cast concrete – a wooden frame is constructed around the stairwell and concrete is poured into the frame, forming the staircase
- pre-cast concrete – concrete is cast in large moulds to form the staircase; usually found in the stairwells of blocks of flats and in other areas of heavy use
- steel – usually found on the exterior of buildings in the form of fire escapes, etc.

Frames and linings

Frames and linings are fitted around openings and are used to allow components such as windows and doors to be fitted. The frame or lining is fitted to the wall and usually finished flush with the walls; the joint between the frame or lining and the wall is covered by the architrave.

Doors

The main purpose of a door is to provide access from one room to another, and to allow a space to be closed off for security/thermal/sound reasons.

Doors come in many varieties, shapes and sizes; the type you need will be determined by the location and purpose of the door. Exterior doors are generally thicker and are fitted with more ironmongery such as letter plates and locks. Some interior doors will have locks fitted as well such as bathroom doors or doors that need to be secure.

Doors can be solid timber or have glass in them and may be graded for fire resistance.

Windows

Windows are fitted to allow natural light to enter the building with minimal loss of heat. Again, windows come in a variety of shapes and styles. Glass that is fitted in a window can be decorative and heat-loss resistant.

Architraves and skirting

Architraves are decorative mouldings used to hide the gap between frames and the wall finish. Skirting is moulding that

Key term

Profile – the shape of a moulding when you cut through it

covers the gap between the floor and the base of a wall. These mouldings come in a variety of **profiles** such as torus and ogee.

Other mouldings can also be used such as picture and dado rails.

Substance damage to building materials

There are a number of substances which can have a negative effect on building materials when they come into contact with them.

Water

Water can cause a range of problems for materials and can have a number of detrimental effects on them. Water can cause components to expand, which can add additional stress to joints, etc. and cause cracks or gaps to appear. Water can also cause some components to rust, reduce the effectiveness of insulation properties and damage interior finishes such as paint. Water in materials will increase the chances that they will be attacked by mould, fungi or insects.

Frost

Frost will have a similar effect to water once it has thawed but during the freezing process materials will shrink with the cold. When the temperature rises and the ice begins to thaw and melt, this change can cause cracks. With some materials this can cause severe problems, particularly with copper plumbing pipes.

Chemicals

Chemicals can have a detrimental effect on building materials. Most will cause some damage, and certain chemicals can corrode or even completely break down and destroy some materials.

Heat and fire

Heat will cause materials to expand, which can cause minor problems such as doors not shutting correctly. It can also affect things such as paint finishes, the thermal properties of materials and the effectiveness of certain glues. Fire will destroy the majority of building materials, including steel, if hot enough and left to burn.

Reducing the risk of damage

Materials can be treated to prevent or reduce the effects of these elements. The type of treatment used is usually chemical and care needs to be taken to ensure that the treatment will not cause other damage to the material. Don't make matters worse when you are trying to improve them!

Find out

Use the Internet to research some common chemicals and the effects that they can have on a range of materials

Products that provide protection to materials will come with manufacturers' instructions – be sure to follow these instructions.

Rectifying material deterioration

If any materials have deteriorated and need to be repaired or replaced, it is vital that the cause of deterioration is found and steps are taken to prevent reoccurrence.

This may mean treating the replacement materials or, in the case of water damage, repairing any leaks. The main methods used to protect and repair the main groups of materials are:

- Timber can be protected by cutting out damaged parts. A protective coating, based on water, tar or solvents, can be placed on wood. These products are known as **preservatives**.
- Metal materials can be given protective coatings. Steel and iron can be coated in zinc, which will prevent rusting. This is known as **galvanic protection**.
- Concrete and masonry are protected by expansion joints, which are designed to safely absorb expansion, contraction and vibration, and which allow movement.

K3. Storage and delivery of materials

Stock rotation and delivery times

When dealing with certain materials it is important to know about stock rotation, which ensures that the materials do not go past their use-by date. Materials such as plaster or cement have a use-by date on them; generally, such materials will set or go off about this date. To prevent this, it is vital that the materials are used before this happens.

When taking delivery of materials, place the newest materials at the back. This will mean that the older materials are used first and will reduce the risk of materials reaching their use-by date.

Delivery dates are also important – you want to ensure that materials are there when they need to be used, but you don't want them delivered weeks before they are needed as they will take up valuable storage space and can get damaged or expire.

The main types of materials affected by this are:

- cement
- plaster
- glue
- paints
- preservative coatings

Checking deliveries to sites

When a delivery is made to a site, security will check that the delivery is due and that the materials are indeed for that site. The foreman or site agent will then look at the delivery note and check it against the order to ensure that what is being delivered is what was ordered.

The unloading of the materials can then take place, usually with a designated person checking the quality of the materials as well as the quantity against the delivery note. The materials should then be stored appropriately, as described above.

Tools used to transport materials

When handling materials it is important to know the best way to carry things safely. Manual handling should be avoided if at all possible and mechanised equipment such as forklifts should be used instead.

If this is not possible or available, hand tools should be used. These include:

- wheelbarrows
- pallet trucks
- bag trolleys
- skips

In some cases there is only one way that a material can be carried, e.g. mixed cement should only be moved in a wheelbarrow.

FAQ

How do I know if the materials I am using are strong enough to carry the load?

The specification will give you the details of the sizes and types of material that are to be used. You will need to use this document when you need to know which materials to use.

Do I have to fix battens to a wall before I plasterboard it?

No, the method called dot and dab can be used where plaster is dabbed onto the back of the plasterboard and then pushed onto the wall.

Check it out

1 Describe the process that should happen before any construction work begins on a building project.
2 State three key factors that influence the design of a foundation and explain why.
3 Explain what is meant by the terms 'dead load' and 'imposed load'.
4 During a site investigation, certain data needs to be collected. Give a list of the key information that must be recorded during this investigation.
5 Why must excavation work be carefully planned before it is carried out?
6 Name three categories of soil.
7 Name three types of foundation. Complete sketches to show the key features of these types of foundation.
8 Explain how surface water can affect excavation work.
9 In a stepped foundation, what is the recommended maximum height of each concrete step?
10 State four of the main principles of building and briefly explain what each of them are.
11 List the three main types of stress.
12 Give a brief description of external walling.
13 What are the four main secondary elements? Why are they secondary?
14 Give a brief description of the process involved with lath and plaster.
15 What are the three main services?

Getting ready for assessment

The information contained in this unit, as well as continued practical assignments that you will carry out in your college or training centre, will help you with preparing for both your end of unit test and the diploma multiple-choice test. It will also aid you in preparing for the work that is required for the synoptic practical assignments.

The information in this unit will build on the information that you may have acquired during Level 1, Unit 1003, and will help you understand the basics of your own trade as well as basic information on several other trade areas.

You will need to be familiar with:

- the principles behind foundations, walls, floor and roofs
- the principles behind internal work
- storage and delivery of building materials.

It is important to understand what other trades do in relation to you and how their work affects you and your work. It is also good to know how the different components of a building are constructed and how these tie in with the tasks that you carry out. You must always remember that there are a number of tasks being carried out on a building site at all times, and many of these will not be connected to the work you are doing. It is useful to remember the communication skills you learnt in Unit 2002, as these will be important for working with other trades on site. You will also need to be familiar with specifications and contract documents to know the type of construction work other crafts will be doing around you on site.

For learning outcome one, you saw the range of different structures, and how they maintain structural stability and quality of insulation. It is important that working drawings are precise in order to complete the structure accurately. You will need to be able to sketch a section through building elements and components.

This unit has explained the different construction methods for foundations, walls, floors and roofs. Although you will not be working on all these elements, you need to be familiar with the work undertaken on them in order to plan when to carry out your own work. You will need to be able to complete a programme of work for a simple two-storey construction. To do this you will need to understand the jobs that other trades have to carry out on these parts of the building.

Remember, a sound knowledge of construction methods and materials will be very useful during your training as well as in later life in your professional career.

Good luck!

CHECK YOUR KNOWLEDGE

1 What is the most popular and widely used insulation?
   a polystyrene sheets.
   b rigid foam.
   c glass mineral wool.
   d rock wool.

2 The purpose of any foundation of a building is to ensure that all dead and imposed loads are safely absorbed and transmitted through to the natural foundation. What is meant by a 'dead load'?
   a the weight of the structure.
   b the weight that may be imposed on the structure.
   c the combined weight of the structure and other loads imposed on it.
   d the weight of the concrete foundation imposed on the natural subsoil or natural foundation.

3 The type of foundation used for a small domestic dwelling or low rise structure is:
   a wide strip foundation.
   b short bored piled foundation.
   c pad foundation.
   d narrow strip foundation.

4 Where a stepped foundation is used, the height of each step should not be greater than the thickness of the concrete and no greater than:
   a 250 mm.
   b 300 mm.
   c 450 mm.
   d 600 mm.

5 In a stepped foundation, the overlap of the concrete to that below should not be less than:
   a 250 mm.
   b 300 mm.
   c 450 mm.
   d 600 mm.

6 Which of the following are primary elements?
   a roofs.
   b windows.
   c foundations.
   d all of the above.

7 Flat roofs are constructed in a similar way to:
   a suspended timber floors.
   b floating floors.
   c truss roofing.
   d cut roofing.

8 Which of the following secondary elements provide access between rooms?
   a skirting.
   b doors.
   c windows.
   d linings.

9 Which of the following are finishing elements?
   a plaster.
   b glazing.
   c dado rails.
   d all of the above.

10 Which of the following can be water- or solvent-based?
   a paint.
   b plaster.
   c wallpaper.
   d render.

Know how to carry out first fixing operations

Carpenters and joiners will undertake many different types of work on site. One of the main areas, and probably the most important, is called 'first fixing'. First fixing is the name given to the work carried out before plastering takes place, and includes studwork, ground lats, stairs, windows and doors. This unit also supports NVQ Unit VR09 Install First Fixing Components.

This unit contains material that supports TAP unit 2 Install First Fixing Components. It also contains material that supports the delivery of the five generic units.

This unit will cover the following learning outcomes:

■ How to fix frames and linings

■ How to fit and fix floor coverings and flat roof decking

■ How to erect timber stud partitions

■ How to assemble, erect and fix straight flights of stairs including handrails.

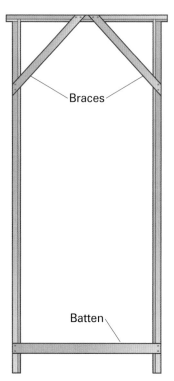

Figure 8.1 Door lining

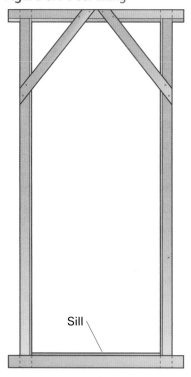

Figure 8.2 Door frame

> **Remember**
>
> Extra care taken with fixing a frame or lining will prevent difficulty later when hanging the door

K1. Know how to fix frames and linings

A key aspect of first fixing is knowing how to fix frames and linings. In this section you will learn the skills needed to work with door frames and linings and window frames and linings. Knowing how to identify and select the correct timber is also covered in this section. A good carpenter or joiner knows the botanical differences between different types of wood and can spot defects – natural or otherwise – in the materials he or she chooses to work with.

Tools for first fixing

As jobs vary, it isn't possible to give a comprehensive list of the appropriate tools for each of the jobs covered in this unit. Methods of work vary from region to region and even from job to job. For example, to fit a window on one job you may require a hammer drill, but another window job will better suit a different percussive tool. Some carpenters prefer to use chisels to cut out a door hinge, while others may use a router. You can never know what problems might arise over the course of a job. The best equipment you can have is a well-stocked toolkit, which should be taken on all jobs. Taking the time to look after your tools is one of the biggest investments you can make. Tools that are looked after will not go missing and are less likely to need replacing.

Door frames and linings

The timber frames to which doors are hung may be divided into two main groups:

- linings (or casings)
- frames.

Door frames are usually constructed from heavier material with solid rebates and are generally used for external entrance doors. Linings are lighter with planted door stops and are generally used for internal doors.

This section will look at doors under the following headings:

- door linings or casings
- door frames
- built-in frames
- storey frames.

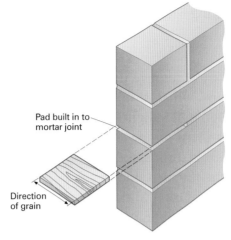

Door linings or casings

Door linings or casings are lightweight internal frames that can be fixed into position by nailing through the jambs into fixing blocks, pads, plugs, or directly into a timber stud wall.

Blocks, or pads, are usually built into the door opening during construction by the bricklayer. Three or four will be needed on each jamb.

Where fixing blocks or pads have not been used, it may be necessary to plug the opening, using a plugging chisel to chase out four mortar joints on each side of the frame. Wooden plugs can be cut from scrap timber, driven into these slots and sawn off plumb. The sawn ends must be kept exactly in line and square. See Figure 8.3.

The door opening should have been built with a clearance of about 20 mm, so that the lining fits loosely and can easily be plumbed and aligned with the walls. Packing will be needed between lining and brickwork so that it is held securely in place once fitted.

The lining can now be offered into the opening and checked that it is plumb and level. It is normal practice to fix one jamb first, making sure that it is plumb and straight and will suit the line of the finished plasterwork. Once this is done the second jamb can be fixed and sighted through to make certain that it is exactly parallel.

It is good practice to make a temporary fixing at the top and bottom of each jamb, leaving the fixing protruding. The lining can then be checked for accuracy and position and adjustments can be made before it is finally fixed. See Figure 8.4.

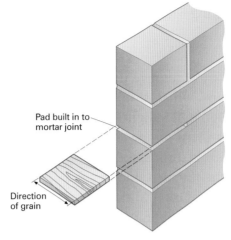

Figure 8.3 Fixing blocks

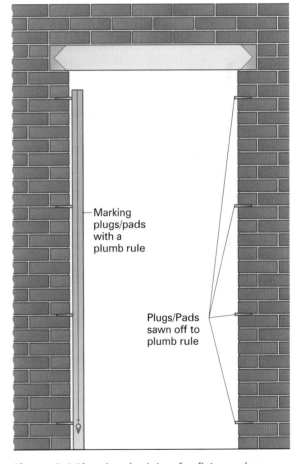

Figure 8.4 Plugging the joints for fixing a door lining

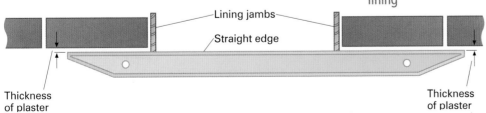

Figure 8.5 Checking alignment of a door lining

The following sequence is one of a possible range of methods.

Step 1 First remove the doorstops, which should have been nailed lightly and put them to one side. The door is unfinished, so check the finished floor level (FFL) in relation to the bottom of the lining. The most accurate way of doing this is to measure down from a datum line (usually 1 m or 900 mm above FFL) and pack the lining to suit.

Step 2 Wedge the lining roughly in position by putting a wedge above each leg in the gap between the lintel and the head of the lining. The head should be checked with a spirit level to see that it is level. The lining can now be packed down the sides with either hardboard or plywood at the top fixing positions. The top of the lining should be moved to ensure equal projection either side of the wall.

Step 3 Fix the lining through the packing at the top with two nails at each fixing point. An alternative is to use screws, which has become common practice.

Step 4 Plumb the lining on the face side and edge using a level. Pack the bottom position as required and fix through one set of packing. Check that the fixed leg is square. Then complete the fixings on the same leg. The amount of fixing points should be around five, but not less than four.

Figure 8.6 Step 1 Measure down

Figure 8.7 Step 2 Wedge the lining

Figure 8.8 Step 3 Fix the lining

Figure 8.9 Step 5 Mark the inside width

Figure 8.10 Step 6 Fit the pinch rod

Step 5 Next, remove the stretcher from the bottom of the lining, hold it at the head and mark the inside width to make a pinch rod.

Step 6 The pinch rod can now be fitted in the bottom position of the lining and the lining packed to suit.

Step 7 Check the leg to see that it is plumb and check the alignment of the frame. The bottom can be fixed, then – moving the pinch rod to ensure the frame is parallel – fix the intermediate positions. The head will only require fixing if the opening exceeds normal width. All nails should be punched to about 3 mm below the surface. The door stops can now be replaced and protection strips fixed if required.

Doorframes

Frames can be fixed to timber stud partitions using 75 mm oval nails. Alternatively, they can be screwed to solid walls using plastic plugs.

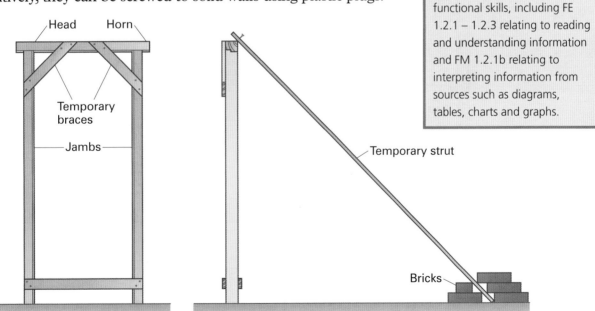

Figure 8.11 Temporary bracing and support

Built-in frames

The majority of frames are 'built-in' by the bricklayer as the brickwork proceeds. Prior to this the frame has to be accurately positioned, plumbed, levelled and struts temporarily inserted by the joiner.

Storey frames

Storey frames are designed to give stability when doorways or fanlights are placed in thin non-loading walls. They are known as storey frames because their height is from floor to floor (one storey). They are used to borrow light from one room to another.

Fire-resistant frames

Fire-resistant frames and frames for fire doors are different to other frames in that they have an intumescent strip which is rebated into the frame. This can help prevent the spread of fire as it expands when heated and seals the gap between frame and door.

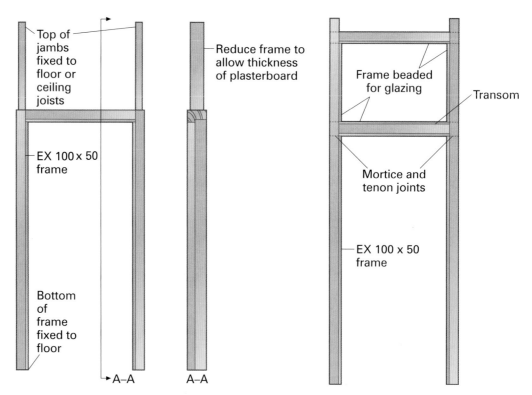

Figure 8.12 Plain storey frame

Figure 8.13 Fanlight storey frame

Window frames and linings

Windows are generally designed to allow daylight and air into a room and give people an outside view, but they must also conserve heat and be able to withstand the weather.

There are many forms and designs, the most common being:

- traditional casement windows
- stormproof windows
- boxed frame sliding sash windows
- bay windows.

Windows are traditionally made from timber, softwood and hardwood but, with new materials and improved manufacturing methods, they are also available in metal, aluminium and pvc (sold under a variety of names including PVC, PVCu, uPVC, PVC-Upvc and others).

As windows are subjected to extreme weather conditions they must be constructed to give maximum protection against the worst possible conditions.

A window should be secure and durable, with easy opening operation, and provide, where necessary, good sound and thermal insulation.

This section will look at:

- the main window types
- the assembly of frames and sashes
- installing windows
- window boards.

Window parts

Different types of windows have different parts, but the majority will have:

- sill – the bottom horizontal member of the frame
- head – the top horizontal member of the frame
- jambs – the outside vertical members of the frame
- mullion – intermediate vertical member between the head and sill
- transom – intermediate horizontal member between the jambs.

If a window has a sash, the sash parts will consist of:

- top rail – the top horizontal part of the sash frame
- bottom rail – the bottom horizontal part of the sash frame
- stiles – the outer vertical parts of the sash frame
- glazing bars – intermediate horizontal and vertical members of the sash frame.

Did you know?

PVC stands for polyvinyl chloride, a tough, white solid plastic that is easy to colour and is strongly resistant to fire, chemicals and the weather

Did you know?

In 1696 houses were taxed on the number of windows per household. This was called 'Window Tax' and it is the reason some very old houses don't have many windows!

Main window types

Traditional casement windows

Casement windows comprise a solid outer frame with one, or more, smaller and lighter frames within it, called sashes or casements.

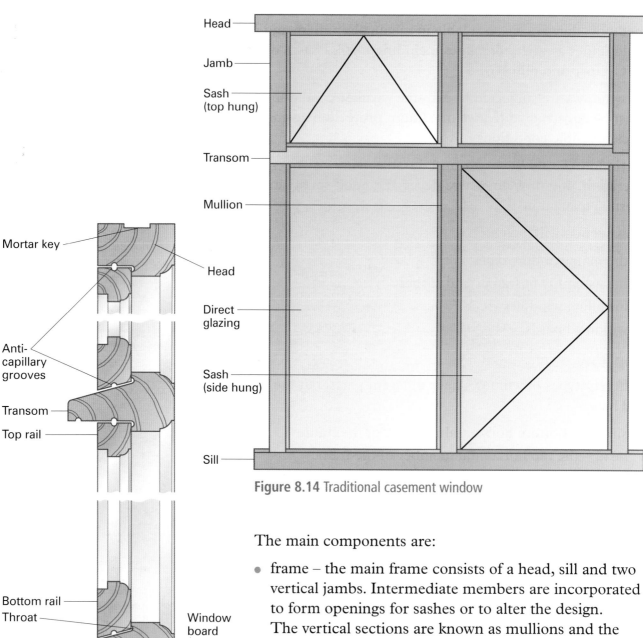

Figure 8.14 Traditional casement window

Figure 8.15 Section through traditional window incorporating top and side hung sashes

The main components are:

- frame – the main frame consists of a head, sill and two vertical jambs. Intermediate members are incorporated to form openings for sashes or to alter the design. The vertical sections are known as mullions and the horizontal members are transoms
- opening casement – consists of top and bottom rails with two stiles. If a casement is divided up, these members are known as glazing bars. When glass is fixed into the main frame itself, this is known as direct glazing

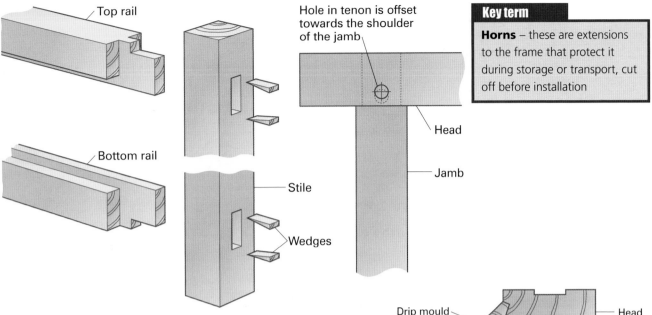

Figure 8.16 Haunched mortise and tenon joint with draw pin

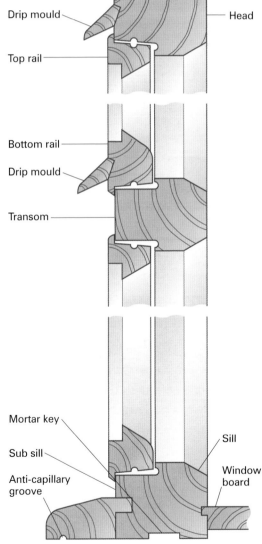

Figure 8.17 Section through typical stormproof window

- joints – all joints used on a traditional casement frame are mortise and tenon, the mortises being formed in the head and sill, while the jambs are tenoned. Casements are haunched mortise and tenons and both are held together using wedges or draw pins and star dowels. Draw pins are more suitable when the **horns** are to be cut off later. Joints are covered in more detail in Unit 2034.

- weather proofing – during manufacture a groove is formed around the outside edges of the head, sill and jambs to provide a mortar key. Grooves, known as anti-capillary grooves, are also incorporated along the inside of the rebates in order to prevent water passing into the building. The transom and sill have a 'throat' to stop water penetrating and are sloped to allow rainwater to run off.

Stormproof windows

Stormproof windows are a variation on traditional casement windows and are designed to increase their performance against bad weather.

The outer frame is basically the same, but the sashes are rebated to cover the gap between sash and frame. This reduces the possibility of driving rain entering the building.

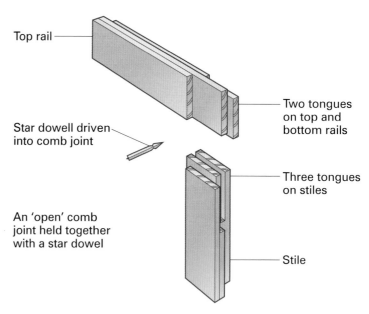

Top rail

Star dowel driven into comb joint

An 'open' comb joint held together with a star dowel

Two tongues on top and bottom rails

Three tongues on stiles

Stile

Figure 8.18 Joining of main frame

The main frame is joined together using mortise and tenon joints, but the sashes are held together using a comb joint and star dowels, although mortise and tenons can be used.

During manufacture anti-capillary grooves are incorporated along the inside of the rebates to prevent water passing into a building.

Assembly of frames and sashes

Windows may be delivered fully constructed. However, if they have to be assembled, the procedure below shows how this should be done:

Step 1 Prior to gluing, frames and sashes should be assembled dry to check the joints, sizes and that they are square.

Step 2 Lay a **squaring rod** across the diagonals to check that each one measures the same. The length of a diagonal should be marked on the rod, then do the same for the other diagonal. If the pencil mark is in the same place the frame is square. New joints may have to be made if the frame is distorted.

Step 3 Once a sash or frame has been assembled dry and checked, a waterproof adhesive can be applied to the faces of the tenons and shoulders and the items glued and cramped together to dry.

Step 4 Check again that the sash is square and not in wind, as well as ensuring the overall size is correct. Then wedges should be lightly tapped into each tenon joint and, if everything is correct, they can be driven in fully, outside wedges first, and star dowels inserted if required.

> **Key term**
>
> **Squaring rod** – any piece of straight timber long enough to lie across the diagonal

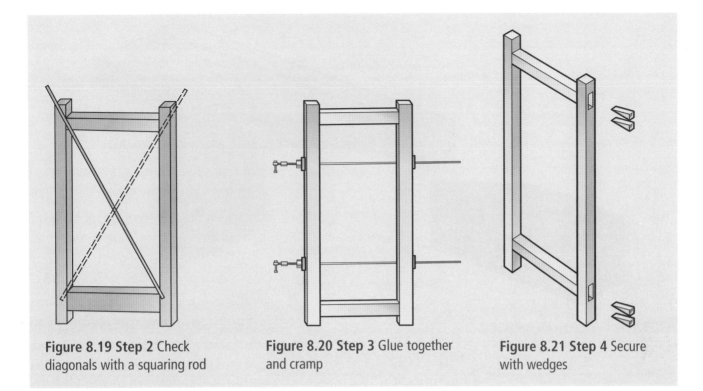

Figure 8.19 Step 2 Check diagonals with a squaring rod

Figure 8.20 Step 3 Glue together and cramp

Figure 8.21 Step 4 Secure with wedges

Installing windows

As brickwork rises

Casement windows made of wood are usually built-in as the brickwork proceeds. They are secured with separate fixing devices, traditionally referred to as frame clamps.

Figure 8.22 Sash clamp

According to the type used, they are either screwed or hammered into the wooden side-jambs as the brickwork rises, where they will be bedded into the mortar between the bricks. Two or three per side is usual, like the built-in door frames covered later.

Initially windows are put in position, checked for plumb and then supported at the head with one or two weighted scaffold boards.

If the windows have a separate sill of stone, or pre-cast concrete, they must be bedded first and protected with temporary boards on their outer face, sides and edges.

Projecting sills, formed with sloping **brick-on-edge**, are usually built at a later stage, the windows having been packed up with spare timber to leave space for these to be inserted.

Key term

Brick-on-edge – a way of laying bricks to form an attractive, projecting window or door sill

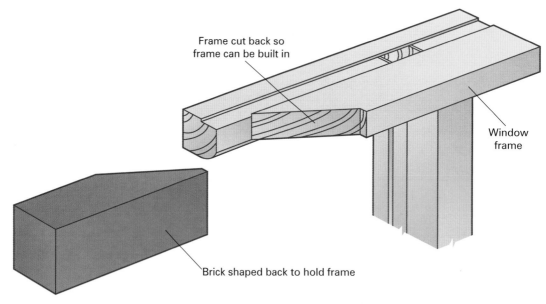

Frame cut back so frame can be built in

Window frame

Brick shaped back to hold frame

Figure 8.23 Built-in window frame

Temporary profile frames

Sometimes it is not possible to build the windows in as the brickwork rises. If this is the case a temporary profile frame may be made on site or in the workshop to match the window size and put in place during building. It can be replaced by the window as soon as the brickwork is dry and stable, using suitable fixings.

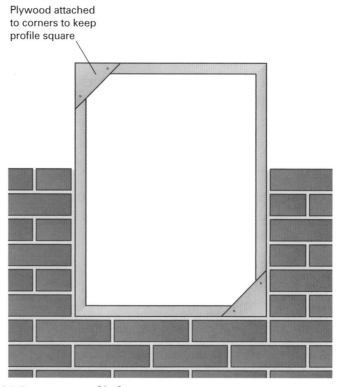

Plywood attached to corners to keep profile square

Figure 8.24 Temporary profile frame

Fixing components

There is a range of commercially available fixing components. Common ones are described below:

- galvanised steel frame cramps – these provide good fixings and, because they are screwed to the frame, any cramps (or ties) already fixed and bedded in mortar are not disturbed by hammering. Also, by resting on the last laid bricks, the next brick above the tie is easily bedded. The disadvantages are their small screws, requiring a screwdriver and bradawl. If there is no groove in the frame, the upturned end of the cramp inhibits the next brick from touching the frame. There will be problems later if rust-proofed screws are not used.

- zinc-plated screw ties – these also provide good fixings and are screwed to the frame, avoiding vibration from hammering. They do not require screws, a bradawl or a screwdriver and can be offset or skewed to avoid the cavities in hollow blocks. On the negative side, the brickwork has to be stopped one course below the required fixing to allow rotation of the loop when screwing in, then the brick or block beneath the tie is bedded – with some difficulty and loss of normal bedding adhesion.

- sherardised holdfasts – holdfasts are fixed quickly and are easily driven in by a hammer. The spiked ends spread outwards when driven into the wood, forming a fishtail with good holding power. The main disadvantage is that the hammering disturbs the frame and permanently loosens any holdfasts already positioned in the **still green** mortar.

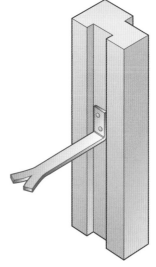

Figure 8.25 Galvanized steel frame cramp

> **Key term**
>
> **Still green** – mortar that is unset and not at full strength

Figure 8.26 Zinc-plated screw tie

Box frame sliding sash window

Often referred to as a box sash window, this is rarely used nowadays – the casement window is preferred as it is easier to manufacture and maintain. Box sash windows are mainly fitted in listed buildings or in buildings where like-for-like replacements are to be used. Sometimes box sash windows are fitted because the client prefers them.

The box sash window is constructed with a frame and two sashes, with the top sash sliding down and the bottom sash sliding up. Traditionally the sashes work via pulleys, with lead weights attached to act as a counterbalance. The weights must be correctly balanced so that the sashes will slide with the minimum effort and stay in the required position. For the top sash, they must be only slightly heavier than the weight of the glazed sash. For the bottom sash, they must be slightly lighter than the weight of the glazed sash.

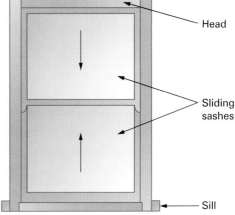

Head

Sliding sashes

Sill

Figure 8.27 Box frame window

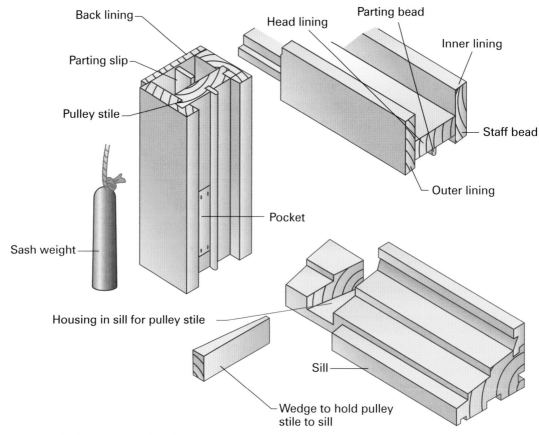

Figure 8.28 Parts of a box sash window frame

Most pulley-controlled systems now use cast-iron weights instead of lead. Newer box sash windows use helical springs instead of the pulley system.

The parts of a box sash window are as follows:

- head lining – the top horizontal member housing the parting bead, to which an inner and outer lining are fixed
- pulley stiles – the vertical members of the frame housing the parting bead, to which inner and outer linings are fixed. The pulley stiles act as a guide for the sliding sashes
- inner and outer linings – the horizontal and vertical boards attached to the head and pulley stiles to form the inner and outer sides of the boxed frame
- back lining – a vertical board attached to the back of the inner and outer linings on the side and enclosing the weights, forming a box
- parting bead – a piece of timber housed into the pulley stiles and the head separating the sliding sashes
- sash weights – cylindrical cast iron weights used to counterbalance the sashes; these are attached to the sashes via

cords or chains running over pulleys housed into the top of the pulley stiles

- sill – the bottom horizontal part of the frame housing the pulley stiles and the vertical inner and outer linings
- staff bead – a piece of timber fixed to the inner lining, keeping the lower sash in place
- parting slip – a piece of timber fixed into the back of the pulley stiles used to keep the sash weights apart
- pockets – openings in the pulley stile that give access to the sash cord and weights, allowing the window to be fitted and maintained.

The sash in a box frame is made slightly differently to normal sashes and has the following parts:

- bottom rail – the bottom horizontal part of the bottom sash
- meeting rails – the top horizontal part of the bottom sash and the bottom horizontal part of the top sash
- top rail – the top horizontal part of the top sash
- stiles – the outer vertical members of the sashes.

Installing a box sash window

This can be done in one go with the sashes and weights already fixed, but for the purposes of this book we will show a more traditional method.

Step 1 First strip the frame down and have the sashes, parting bead, staff bead and pockets removed.

Step 2 Then fix the frame into the opening, ensuring that it is both plumb and level and fit the sash cord and sashes (the fitting of the sash cord to the sashes will be covered in Unit 2011).

Step 3 Next, slide the top sash into place and fit the parting bead, to keep the top sash in place.

Step 4 Now fit the bottom sash, then the staff bead, to keep the bottom sash in place.

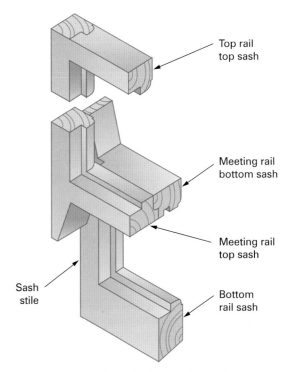

Figure 8.29 Sashes of a box sash window

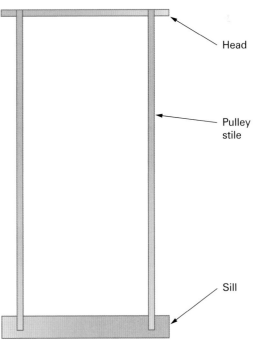

Figure 8.30 Stripped down frame

Unit 2008 Know how to carry out first fixing operations

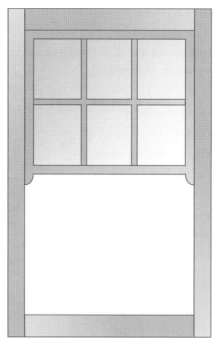

Figure 8.31 Top sash in place

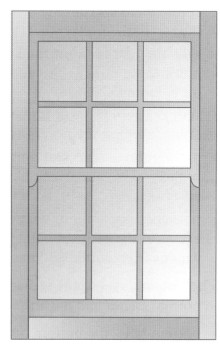

Figure 8.32 Top and bottom sashes in place

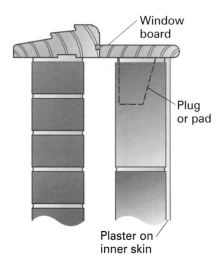

Figure 8.33 Fixed window board

Step 5 Finally, seal the outside with a suitable sealer, fit the ironmongery and make good to the inside of the window.

The marking out of windows will be covered in Unit 2035 and the maintenance of a box sash window in Unit 2011.

Window boards

The window board forms the finish or trim to the top of the inner leaf of the cavity wall, where it finishes at the window opening. Window boards can be formed from solid timber, blockboard, medium density fibreboard (MDF), plywood or plastic. They are used to give a neat finish on the inside of the window sill. They are usually fixed before plastering and are thus part of the first fix.

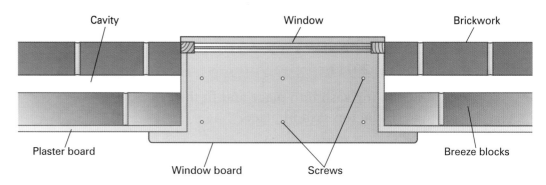

Figure 8.34 Plan view of window board fixed in position

The window board shown in Figure 8.33 has been fixed to wooden plugs (similar to fixing a door lining). The back edge of the window board is fitted into a groove in the sill to allow for any movement in the timber. This will prevent a gap showing should shrinkage occur, which can often happen as radiators are usually situated below the windows.

The front end of the window board is usually nosed or rounded and the ends cut to run past the brick jambs.

The sequence for fixing window boards is as follows:

Step 1 Check the board for length. Then mark and cut to length as necessary.

Step 2 Check for height and level, ensuring engagement of tongue and groove, then make and fix **packings** until height and level are achieved.

Step 3 Fix board with nails through packings, if this can be done into material suitable to receive nails. Otherwise use plugs.

Key term

Packings – any material, generally waste wood, used to fill a gap

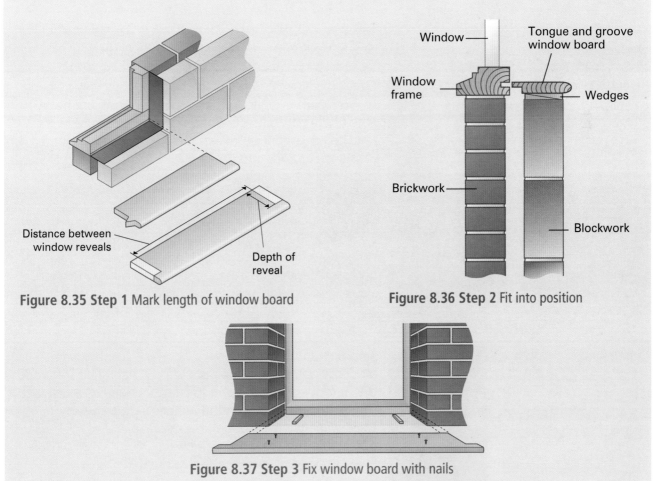

Figure 8.35 **Step 1** Mark length of window board

Figure 8.36 **Step 2** Fit into position

Figure 8.37 **Step 3** Fix window board with nails

Pivot windows

Centre hung or pivot windows are used mainly in high rise buildings or similar areas, as they allow the outside of the window to be cleaned easily from the inside. The window consists of a solid frame into which the casement is fitted and hung on pivot hinges in the centre.

> **Did you know?**
>
> The back edge of a window board can be rebated to form a tongue to fit into a groove on the inside sill of a window

Timber identification and selection

To select the right timber for the job, you must first know how to identify the different varieties of wood you will encounter. This has been covered in detail in Level 1, but the table and figures on pages 132–135 can be used as a handy quick-reference tool.

Timber is divided into the categories hardwood and softwood, although these names do not describe the properties of the timber. Instead these divisions come from botany, and are a scientific classification. The tables are a useful reference to the key properties of some of the more common woods you will use.

Name	Main sources	Identification	Properties/Description	Uses
Douglas Fir	USA, Canada		Straight grained and resilient – easy to work with by hand or machine. One of the hardest softwoods – can take heavy, continuous wear. A high resistance to acids and decay. Has good gluing and high insulation qualities.	First-class joinery, light and heavy structural work, glued laminated work and plywood
Larch	Europe		Very strong and durable, average to work with, and sawing and machine can leave a decent finish, but loose knots can be troublesome.	Fencing, gate posts, garden furniture and railway sleepers
Pitch pine	South USA		Similar strength to Douglas Fir. Works moderately easily, but the resin is often troublesome, tending to clog saw-teeth and cutters and to adhere to machine tables. A good finish is obtainable and the wood can be glued satisfactorily. Takes nails and screws well.	Shipbuilding, polished softwood joinery and church furniture

Name	Main sources	Identification	Properties/Description	Uses
Redwood (commonly known as Pine)	Europe		Moderately strong for its weight with average durability. The timber works easily and well with both hand and machine tools, but ease of working and quality of finish is dependent on the size, the number of knots and amount of resin. The wood is capable of a smooth, clean finish, and can be glued, stained, varnished and painted. Takes nails and screws well.	Depending on quality it can be used for interior or exterior work and can be used for most tasks such as carcassing and finish joinery
Whitewood (also known as European Spruce)	Europe		Similar to redwood or Scots pine in strength and durability. Good to work with by both hand and machine. Takes glue, nail and screws well and can produce a good finish.	Similar uses to redwood
Western red cedar	Canada, USA		Not as strong as redwood, but has naturally occurring oils which prevents insect attack. Non-resinous and straight grained. Good workability with both hand and machine. Does not need treating as the wood will stand up to severe weather and will turn a silvery colour when exposed.	Externally for good quality timber buildings, saunas, etc.

Table 8.1 Identification of softwoods

Name	Main sources	Identification	Properties/Description	Uses
Ash	Europe		Straight grained, very tough and flexible. Although tough, ash works quite well with both hand and machine tools and finishes to a reasonable smooth finish. It can be glued, stained and polished and takes nails and screws well.	Furniture, boat-building, sports equipment, tool handles, etc.
Oak	Europe		Very strong with English oak being the strongest. Good resistance to bending and shearing and workability is quite tough, but can be planed to a good finish. Susceptible to fungi attack and ironwork should not be placed onto untreated oak as it will disfigure the timer and leave a bluish stain. Glues well, but difficult to screw and nail.	High-class joinery, panelling, doors, exposed roofing, etc.

Name	Main sources	Identification	Properties/Description	Uses
Beech	Europe		Hard, close grained and durable with a fine texture. Works fairly well by both hand and machine and is capable of a good, smooth surface. Takes glue, stains and polishes well and can produce an excellent veneer.	Furniture, kitchen utensils, wood block floors, etc.
Mahogany	Africa		Interlocked grain, reasonably strong for its weight and moderately resistant to decay. Fairly easy to work with hand and machine, takes glues, finish, nails and screws well.	High-class joinery, furniture, boat-building and plywood veneers
Mahogany	Cuba		Extremely strong for its weight; it also has an interlocking grain. Very good to work with and takes glue, nails, etc. very well and can be polished to an excellent finish.	The same as African mahogany but considered to be superior
Maple rock	Canada, North east USA		Fine grained with excellent finishing qualities and average durability. Can be difficult to work and is hard to nail or screw.	Panelling, flooring, furniture and snooker cues
Sapele	West Africa		Harder than mahogany with similar strength properties to oak. Works fairly well with hand and machine tools, but the interlocked grain is often troublesome in planing and moulding. It takes screws and nails well, glues satisfactorily, stains readily, and takes an excellent polish.	Furniture, veneers, etc.
Teak	India, Java, Thailand		Greasy to touch and resistant to both insect and fire, the timber is strong, very durable, moderately elastic and hard, but is rather brittle along the grain. Workability can be difficult and tools soon become blunt, but the wood is capable of a good finish if cutting edges are kept sharpened. It can be glued, stained and polished. It holds screws and nails reasonable well, but the wood is inclined to be brittle.	High-class joinery, furniture, boat-building, etc.

Name	Main sources	Identification	Properties/Description	Uses
Walnut	Europe		A relatively tough wood with good resistance to splitting. Easy to work with hand and machine and takes glue, nails and screws well. Can be polished to an excellent finish. Staining is likely if in contact with iron under damp conditions.	Furniture veneers, etc.

Table 8.2 Identification of hardwoods

Timber defects

Defects are faults that are found in timber. Some present a serious structural weakness in the timber, while others do little more than spoil its appearance.

Defects can be divided into two groups:

- seasoning defects (bowing, springing, winding (or twist), cupping, shaking, collapse, case hardening)
- natural defects (heart shakes, cup shakes, star shakes, knots).

Timber defects have been covered in great detail in Level 1. Below you will find illustrations of the major types of defects, both seasoning and natural, to refresh your memory.

Seasoning defects

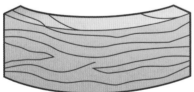

Figure 8.38 Bowing

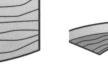

Figure 8.39 Springing

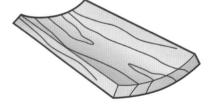

Figure 8.40 Cupping

Figure 8.41 Shaking

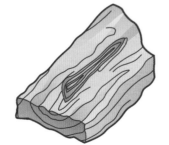

Figure 8.42 Winding (or twist)

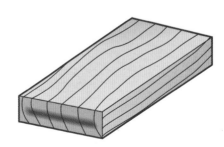

Figure 8.43 Collapse

Natural defects

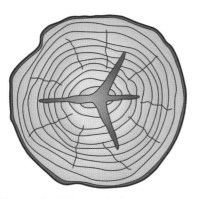

Figure 8.45 Heart shakes

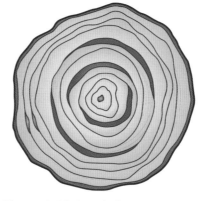

Figure 8.46 Cup shakes

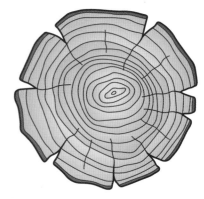

Figure 8.47 Star shakes

Knots

Knots mainly occur in softwood and mark the origin of a branch in the tree. Knots are termed either 'dead' or 'live' depending on the condition of the branch which caused it. Dead knots are often loose. Small live knots are not really a problem. Small dead knots and large knots, either dead or live, are a serious structural weakness.

Types of knot are shown in Figures 8.48 to 8.51.

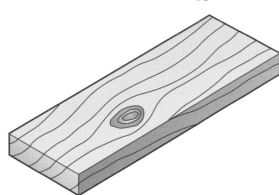

Figure 8.48 Face knot

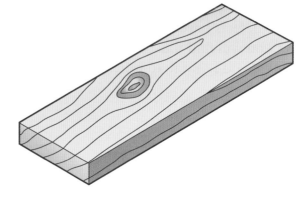

Figure 8.49 Dead knot (often loose)

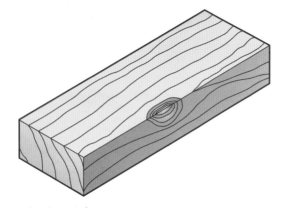

Figure 8.50 Arris knot

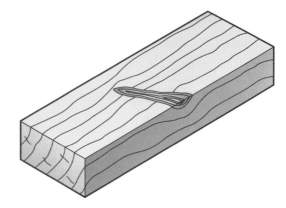

Figure 8.51 Splay knot

The rough unprocessed edge of the tree which still has bark on it is called the wane. A waney edge is the result of economical conversion, where the rough edge is still attached.

Timber decay

The decay of timber is caused by wood-destroying fungi or wood-boring insects – and sometimes both. Fungal decay of timber must be dealt with as soon as it is identified, as both wet and dry rot will attack timber until it is destroyed if the conditions are right.

Dry rot

The main conditions for an attack of dry rot are:

- damp timber, with a moisture content above 20 per cent (known as the **dry rot safety line**)
- poor, or no, ventilation (i.e. no circulation of fresh air).

Identification of dry rot

Dry rot can be identified by:

- an unpleasant, musty smell
- visible distortion of infected timber; warped, sunken and/or shrinkage cracks
- probing to test the timber for softening or crumbling
- the appearance of fruiting bodies
- the presence of fine, orange-red dust on the floorboards and other parts of the structure
- the presence of whitish-grey strands on the surface of the timber.

Eradication of dry rot

Dry rot is eradicated by carrying out the following actions:

- eliminate all possible sources of damp, such as blocked air bricks, bridged damp proof course and leaking pipes.
- determine the extent of the attack
- remove all infected timber
- clean and treat surrounding walls, floors, etc. with a suitable fungicide
- treat any remaining timber with a preservative
- replace rotted timber with new, treated timber
- monitor completed work for signs of further attack.

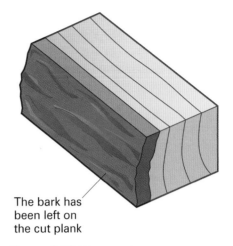

The bark has been left on the cut plank

Figure 8.52 Waney edge

> **Did you know?**
>
> Each season trees grow a new layer of wood, which shows as a ring when we cut the trunk; the thickness varies depending on growing conditions that year. Rings on trees from tropical areas rarely show because there is little seasonal variation

> **Key term**
>
> **Dry rot safety line** – when the moisture content of damp timber reaches 20 per cent. Dry rot is likely if the moisture content exceeds this

> **Did you know?**
>
> Fungal fruiting bodies are the sign of advanced decay in timber

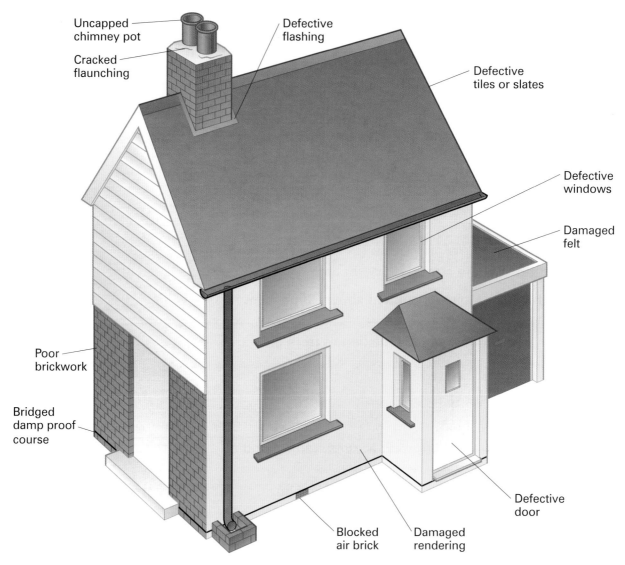

Figure 8.53 Possible causes of dry and wet rot

Remember

Timber affected by dry rot, wet rot or insect attack should be cut back at least 600 mm

Wet rot

This is a general name given to another type of wood-destroying fungus. The conditions where wet rot is found are usually wet rather than damp. Although wet rot is capable of destroying timber it is not as serious a problem as dry rot and, if the source of wetness is found, the wet rot can be halted.

Identification of wet rot

The signs to look for to identify wet rot are:

- timber becomes darker in colour with cracks along the grain
- decay usually occurs internally leaving a thin layer of relatively sound timber on the outside

- localised areas of decay close to wetness
- a musty, damp smell.

Eradication of wet rot

As wet rot is not as serious a problem as dry rot, less extreme measures are normally involved. It is usually sufficient to remove the rotted timber, treat the remaining timber with a fungicide and replace any rotted timber with treated timber. Finally, if possible, the source of any wetness should be rectified.

K2: Know how to fit and fix floor coverings and flat roof decking

As with all forms of carpentry and joinery, it is vital that you select the appropriate materials for the job. When working with floor coverings and decking, durability is key – for example stiletto heels can gouge a floor made from a softer wood. Knowing the different types of timber and other materials that you will be working with throughout your career is as important to your role as spelling and grammar are to an author.

Use the two tables on pages 132–135 to familiarise yourself with some of the more common types of hardwoods and softwoods. Once you understand their properties you will be able to determine which categories other woods not listed belong to.

Flooring

Softwood flooring

Softwood flooring can be used at either ground or upper floor levels. It usually consists of 25 × 150 mm tongued and grooved (T&G) boards. The tongue is slightly off-centre to provide extra wear on the surface that will be walked upon.

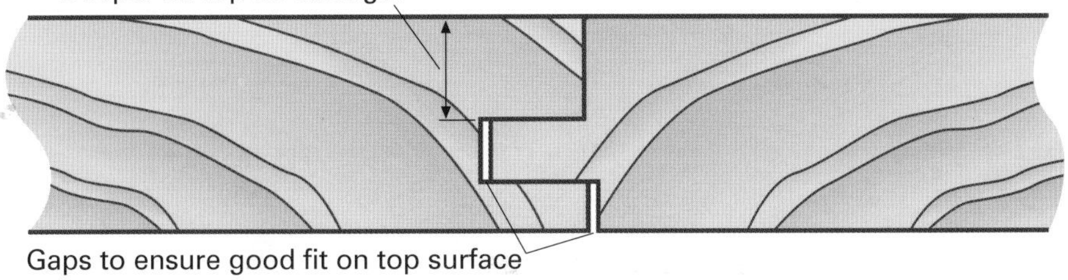

Figure 8.54 Section through softwood covering

Deeper on top for strength

Gaps to ensure good fit on top surface

Functional skills

To select the correct materials for a job you will need to read and use design specifications. When you are reading these specifications, and using the drawings that go with the, you are practicing a number of functional skills including:

FE 1.2.1 - Identifying how the main points and ideas are organised in different texts.

FE 1.2.2 – Understanding different texts in detail.

FE 1.2.3 – Read different texts and take appropriate action, e.g. respond to advice/instructions.

FM 1.2.1b – Interpret information from sources such as diagrams, tables, charts and graphs

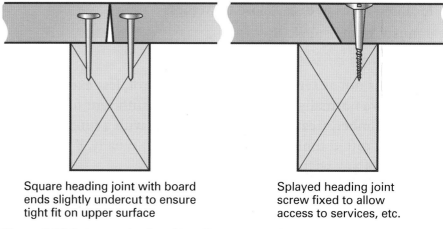

Square heading joint with board ends slightly undercut to ensure tight fit on upper surface

Splayed heading joint screw fixed to allow access to services, etc.

Figure 8.55 Square and splayed heading

When boards are joined together, the joints should be staggered evenly throughout the floor to give it strength. They should never be placed next to each other, as this prevents the joists from being tied together properly. The boards are either fixed with floor brads nailed through the surface and punched below flush, or secret nailed with lost head nails through the tongue. The nails used should be 2½ times the thickness of the floorboard.

The first board is nailed down about 10–12 mm from the wall. The remaining boards can be fixed four to six boards at a time, leaving a 10–12 mm gap around the perimeter to allow for expansion. This gap will eventually be covered by the skirting board.

There are two methods of clamping the boards before fixing:

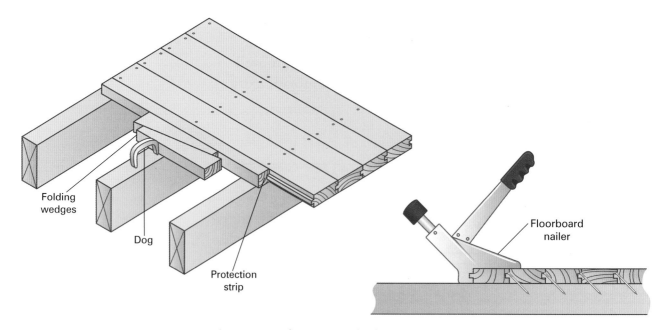

Folding wedges

Dog

Protection strip

Floorboard nailer

Figure 8.56 Clamping methods

Hardwoods can also be used as floor boards. They are more decorative and much more expensive.

Chipboard flooring

Flooring-grade chipboard is increasingly being used for domestic floors. It is available in sheet sizes of 2440 × 600 × 18 mm and can be square-edged or tongued and grooved on all edges, the latter being preferred. If square-edged chipboard is used it must be supported along every joint.

Tongued and grooved boards are laid end-to-end, at right angles to the joists. Cross-joints should be staggered and, as with softwood flooring, expansion gaps of 10–12 mm left around the perimeter. The ends must be supported.

When setting out the floor joists, spacing should be set to avoid any unnecessary wastage. The boards should be glued along all joints and fixed using either 50–65 mm annular ring shank nails or 50–65 mm screws. Access traps must be created in the flooring to allow access to services such as gas and water.

Access traps should ideally be taken into consideration when the floor covering is initially laid and the sheets or boards should be cut so that the traps can be formed. Occasionally you may need access to a pre-laid floor but find that there are no access traps. If this is the case, you will need to create your own.

Access traps can be formed by finding the nearest joists and cutting between them to create an opening. This is done either using a floorboard saw or by drilling a hole and using a jigsaw. Great care must be taken when carrying out this task as there is no real way to tell what services, such as electric cables or water/gas pipes, are underneath.

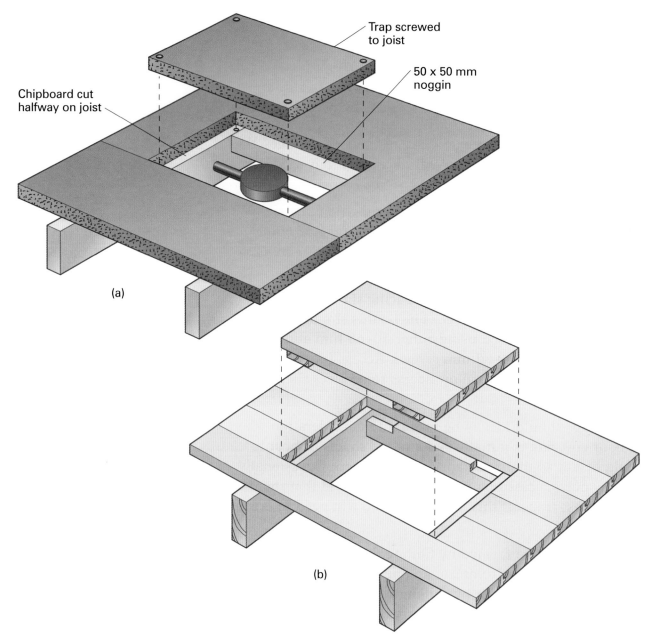

Trap screwed to joist

50 x 50 mm noggin

Chipboard cut halfway on joist

(a)

(b)

Figure 8.57 Access traps, (a) Chipboard floor, (b) Tongued and grooved floor

Flat roofs

A flat roof is any roof which has its upper surface inclined at an angle (also known as the fall, slope or pitch) not exceeding 10 degrees.

A flat roof has a fall to allow rainwater to run off, preventing puddles forming as they can put extra weight on the roof and cause leaks. Flat roofs will eventually leak, so most are guaranteed for only 10 years (every 10 years or so the roof will have to be stripped back and re-covered). Today **fibreglass** flat roofs are available that last much longer, so some companies will give a 25-year guarantee on their roof. Installing a fibreglass roof is a job for specialist roofers.

The amount of fall should be sufficient to clear water away to the outlet pipe(s), or guttering, as quickly as possible across the whole roof surface. This may involve a single direction of fall or several directional changes of fall such as:

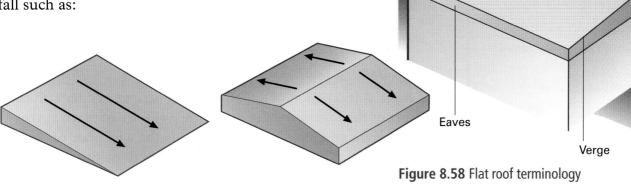

Figure 8.58 Flat roof terminology

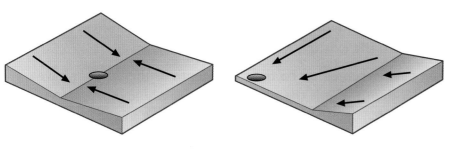

Figure 8.59 Falls on a flat roof

Did you know?

If a puddle forms on a roof and is not cleared away quickly, over a period of years the water will eventually work its way through

Find out

Why are flat roofs more at risk from wet weather than hipped roofs?

Construction of a flat roof

Flat roof joists are similar in construction to floor joists (discussed later in this unit) but unless they are to be accessible, they are not so heavily loaded. Flat roof joists are therefore of a smaller dimension than those used in flooring.

There are many ways to provide a fall on a flat roof. The method you choose depends on the direction of fall and where on the building the roof is situated.

Laying joists to a fall

This method is by far the easiest – all you have to do is ensure that the wall plate fixed to the wall is higher than the wall plate on the opposite wall or vice versa. The problem is that this method will also give the interior of the roof a sloping ceiling. This may be fine for a room such as a garage, but for a room such as a kitchen extension the client might not want the ceiling to be sloped and another method would have to be used.

Joists with firring pieces

Firring pieces provide a fall without disrupting the interior of the room, but involve more work. Firrings can be laid in two different ways, depending on the layout of the joists and the fall.

The layout of joists is explained in more detail in the flooring section of this unit.

Using firring pieces

The basic construction of a flat roof with firrings begins with the building of the exterior walls. Once the walls are in place at the correct height and level, the carpenter fits the wall plate on the eaves wall (there is no need for the wall plate to be fitted to the verge walls). This can be bedded down with cement, or nailed through the joints in the brick or block work with restraining straps fitted for extra strength. The carpenter then fixes the **header** to the existing wall.

The header can either be the same depth as the joists or have a smaller depth. If it is the same depth, the whole of the joist butts up to the header and the joists are fixed using joist hangers. If the header has a smaller depth, the joists can be notched to sit on top of the header, as well as using framing anchors.

Once the wall plate and header are fixed, they are marked out for the joists at the specified centres (300 mm, 400 mm or 600 mm). The joists are cut to length, checked for **camber (crown)** and

Key terms

Firring piece – long wedge tapered at one end and fixed on top of joists to create the fall on a flat roof

Header – a piece of timber bolted to the wall to carry the weight of the joists

Camber (crown) – the bow or round that occurs along some timbers' depth

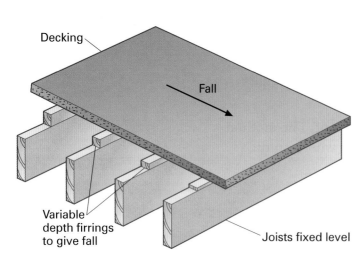

Figure 8.60 Joists with firring pieces

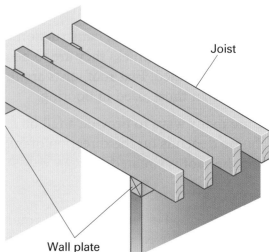

Figure 8.61 Joists laid to a fall

fixed in place using joist hangers. Strutting, or **noggings**, are then fitted to help strengthen the joists.

Once the joists are fixed in place they must have restraining straps fitted. A strap must be fitted to a minimum of one joist per 2 m run, then firmly anchored to the wall to prevent movement in the roof under pressure from high winds.

The next step is to fix the firring pieces, which are either nailed or screwed down onto the top of the joists. Insulation is fitted between the joists, along with a vapour barrier to prevent the movement of moisture caused by condensation.

Decking

Once the insulation and vapour barrier are fitted, it is time to fit the decking. Decking a flat roof can be done with a range of materials including:

- tongued and grooved board – these boards are usually made from pine and are not very moisture-resistant, even when treated, so they are rarely used for decking these days. If used, the boards should be laid either with, or diagonal to, the fall of the roof. Cupping of the boards laid across the fall could cause the roof covering to form hollows in which puddles could form
- plywood – only roofing grade boards stamped with **WBP** should be used. Boards must be supported and securely fixed on all edges in case there is any distortion, which could rip or tear the felt covering. A 1 mm gap must be left between each board along all edges in case there is any movement caused by moisture, which again could cause damage to the felt
- chipboard – only types with the required water resistance classified for this purpose must be used. Boards are available that have a covering of bituminous felt bonded to one surface, giving temporary protection against wet weather. Once laid, the edges and joints can be sealed. Edge support, laying and fixing are similar to floors (covered elsewhere in this unit). Moisture movement will be greater than with plywood as chipboard is more **porous**, so a 2 mm gap should be allowed along all joints, with at least a 10 mm gap around the roof edges. Tongued and grooved chipboard sheets should be fitted as per the manufacturer's instructions
- oriented strand board (OSB) – generally more stable than chipboard, but again only roofing grades must be used. Provision for moisture movement should be made as with chipboard

Key term

Translucent – allowing some light through but not clear like glass

Did you know?

Bitumen is bought as a solid material. It is then broken up, placed in a cauldron and heated until it becomes a liquid, which can be spread over the flat roof. The bitumen will only stay in liquid form for a few minutes before it starts to set. Once set, it forms a waterproof barrier

Key term

Bitumen – also known as pitch or tar, bitumen is a black sticky substance that turns into a liquid when heated and is used on flat roofs to provide a waterproof seal

Safety tip

Bitumen in liquid form is very hot and if it comes in contact with the skin will stick and cause severe burns. When working with bitumen you must wear gloves and goggles and ensure that your arms and legs are fully covered

Did you know?

Flat roofs have stones or chippings placed on top to reflect the heat from the sun, which can melt the bitumen, causing problems with leaks

- cement-bonded chipboard – strong and durable with a high density (much heavier and greater moisture resistance than standard chipboard). Provisions for moisture movement should be the same as chipboard
- metal decking – profiled sheets of aluminium or galvanised steel with a variety of factory-applied colour coatings and co-ordinated fixings are available. Metal decking is more usually associated with large steel substructures and fixed by specialist installers, but it can be used on small roof spans to some effect. Sheets can be rolled to different profiles and cut to any reasonable length to suit individual requirements
- translucent sheeting – this might be corrugated or flat (e.g. polycarbonate twin wall) and must be installed as per the manufacturer's instructions.

Metal decking and **translucent** sheeting are supplied as finished products, but timber-based decking needs additional work to it to make it watertight, as explained in the next section.

Weatherproofing a flat roof

Once a flat roof has been constructed and decked the next step is to make it watertight. The roof decking material can be covered with different materials using different methods. One basic way of covering the decking is as follows.

The roof decking is covered in a layer of hot **bitumen** to seal any gaps in the joints. Another layer of bitumen is then poured over the top and felt is rolled onto it, sticking fast when the bitumen sets. A second roll of felt is stuck down with bitumen, but is laid at 90 degrees to the first. Some people add more felt at this stage – sometimes up to five more layers.

The final step can be done in a number of ways. Some people put stones or chippings down on top of the final layer; others use felt that has stones or chippings embedded and a layer of dried bitumen on the back. The felt with stones embedded into it is laid by rolling the felt out and using a gas blowtorch to heat the back, which softens the bitumen allowing it to stick.

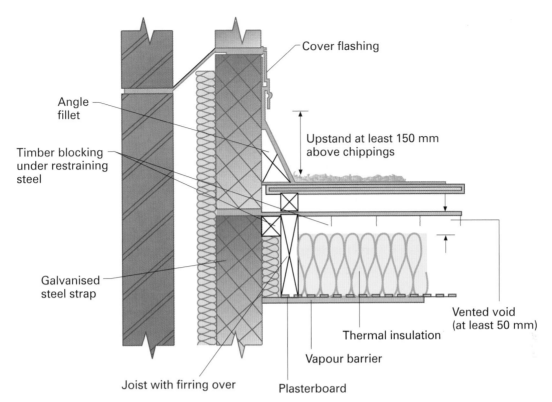

Cover flashing

Angle fillet

Timber blocking under restraining steel

Upstand at least 150 mm above chippings

Galvanised steel strap

Vented void (at least 50 mm)

Thermal insulation

Vapour barrier

Joist with firring over

Plasterboard

Figure 8.62 Abutment joint

Finishing a flat roof at the abutment and verge

Abutment

The abutment finish needs to take into consideration the existing wall, as well as the flat roof. The abutment is finished by cutting a slot into the brickwork, or blockwork, and fixing lead to give a waterproof seal, which prevents water running down the face of the wall and into the room. **Tilt** or **angle fillets** are used to help with the run of the water and to give a less severe angle for the lead to be dressed to.

Verge

Since a flat roof has such a shallow pitch, the verge needs some form of upstand to stop the water flowing over the sides of the verge instead of into the guttering at the eaves. This is done using tilt, or angle fillets, nailed to the decking down the full length of the verge prior to the roof being felted or finished.

Eaves details

The eaves details can be finished in the same way as the eaves on a pitched roof, with the soffit fitted to the underside of the joists and the fascia fitted to the ends.

Did you know?

Dressing lead means moulding the lead so that it covers or fits into something. Lead dressing is usually done by plumbers and roofers

Key term

Tilt or **angle fillet** – a triangular piece of wood used in flat roof construction

K3. Know how to erect timber stud partitions

Timber studwork covers partitions or walls, usually of light construction, used to divide a building or large area into compartments. The NVQ syllabus only covers non-load-bearing walls.

This section will cover:

- basic construction
- in-situ partitions
- pre-made partitions
- holes and notches
- insulation.

Basic construction

Timber partitions are made up of a head and sole plate with studs at regular centres fixed between them. These are stiffened up with noggings, which also provide extra fixings for sheet material, usually plasterboard, and fixing points for heavy components such as wash basins, toilets. See Figure 8.63.

The timber used is normally 75 mm × 50 mm for heights up to 2.4 m and 100 mm × 50 mm for studwork above this height. Planed all round timber is preferred because its cross-section is uniform and is better to handle.

Find out

When were stud partitions introduced to the building trade? What was in use before?

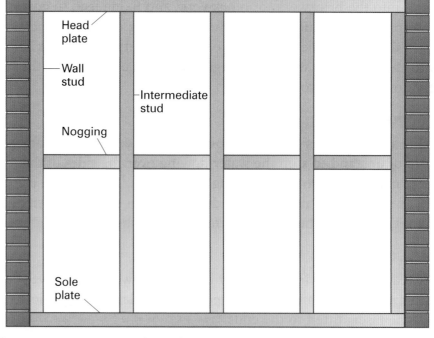

Figure 8.63 Components of a stud partition

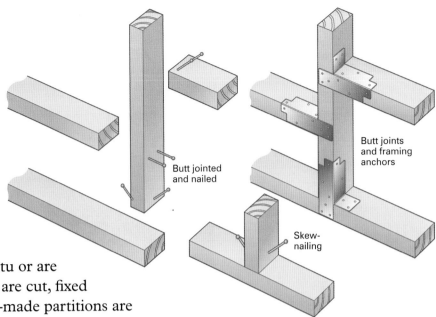

Butt jointed
and nailed

Butt joints
and framing
anchors

Skew-
nailing

Figure 8.64 Methods of fixing studwork together

Partitions are either made in-situ or are pre-made. In-situ components are cut, fixed together and fitted on site. Pre-made partitions are assembled on site or in a factory, for erection on site later.

The intermediate studs are measured and fixed to suit the sheet material to be fixed. For 9.5 mm, plasterboard studs are spaced at 400 mm centre to centre and for 12 mm plasterboard at 600 mm. These sheets should cover 2.4 m × 1.2 m. A sheet size of 2.4 m × 0.9 m would have studs spaced at 450 mm.

The majority of stud partitions are fixed together by butt joints and skew-nailed. Another method is to use framing anchors, which are strong and quick to fit.

In-situ partitions

Fixing partitions in position

Step 1 Mark out positions for partitions on the floor. Check the ceiling to see how to fix the head plate. Ideally it will run at a right angle to the joists but, if not, further noggings should be fixed between each joist to provide a fixing point. The head plate can now be fixed to the ceiling, usually with 100 mm wire nails or screws.

Figure 8.65 Step 1 Mark out positions

Figure 8.66 Step 2 Plumb down

Figure 8.67 Step 3 Fix the sole plate

Figure 8.68 Step 4 Fix studs

Figure 8.69 Step 5 Position intermediate studs

Figure 8.70 Step 6 Skew-nail the studs

Figure 8.71 Step 7 Through-nail central noggings

Figure 8.72 Step 7 Skew-nail the studs

Step 2 Plumb down from the head plate with a plumb bob, or a level and straight edge, and mark the position on the floor.

Step 3 The sole plate can be fixed in the same way as the head plate. If on a concrete floor, it should be plugged and screwed.

Step 4 Cut the wall studs and fix to the wall by plugs and screws. They should be skew-nailed to the head and sole plate.

Step 5 Mark the position of intermediate studs on the side of the head and sole plates at the appropriate centres, that is 400 mm, 450 mm or 600 mm depending on sheet material size and thickness.

Step 6 Measure each stud individually, as they may vary in length between head and sole plate, then skew-nail with 100 mm wire nails. Studs should be a tight fit and, with the bottom in place, the tip should be forced over into position.

Step 7 If overall partition height is greater than 2.4 m, position noggings centrally at the top edge of the boards. Through-nail one end and skew-nail the other.

Step 8 Fix additional noggings at various heights from 600 mm, 1200 mm to 1800 mm to give extra strength and rigidity to the frame, and provide extra fixing points for the sheet material and any components that may have to be fixed to the walls.

Figure 8.73 Step 8 Fix additional noggings

Figure 8.74 Noggings provide fixings for sheet materials

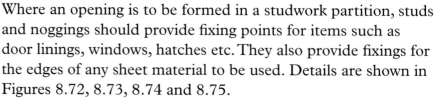

Figure 8.75 Section through a door opening

Forming openings

Where an opening is to be formed in a studwork partition, studs and noggings should provide fixing points for items such as door linings, windows, hatches etc. They also provide fixings for the edges of any sheet material to be used. Details are shown in Figures 8.72, 8.73, 8.74 and 8.75.

Corners and junctions

When a partition has to be returned at right angles or when a 'tee' junction is required, extra studs must be fixed to provide support and fixing for the covering material.

Figure 8.76 Opening formed to take a window lining

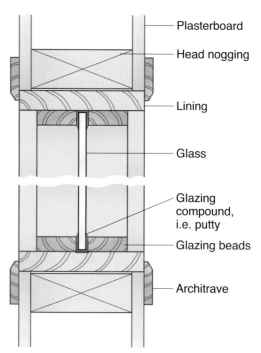

Figure 8.77 Section through a window lining

Figure 8.78 Corner detail with extra studs

Figure 8.79 T-junction detail with extra studs

Pre-made partitions

Pre-made partitions are the same specification as in-situ partitions, but can be made up either in a workshop or on site. They are made slightly under size in height and width to allow for the frame to be offered into position.

Joints are usually butt-jointed and nailed, but can also be housed-in or framing anchors used.

Folding wedges can be used to keep the frame in position while fixing takes place.

Holes and notches

It is common practice for pipes and cables to be concealed within timber partitions, though it is recommended that they are kept to a minimum so as not to affect partition strength.

Figure 8.80 Holes for pipework

Figure 8.81 Notches with wire

Positioning is important. Holes for carrying cables, and notches for water pipes, should be kept away from areas where there is a possibility they may be punctured by a nail or screw. This includes areas where, for example, kitchen units, cupboards, skirting boards, dado or picture rails may be fixed.

Notches can be protected by fixing a metal plate over them.

Insulation

Certain situations may require studwork to have insulation between the wall coverings to form a sound or thermal barrier. *Building Regulations 2000* control the methods of achieving this and should be referred to for technical information.

Extra fire-resistance can be achieved by double boarding the studwork or by using fire-lined board.

Mineral wool, or glass fibre insulation, are most commonly used for sound and thermal insulation.

Figure 8.82 Fire-resistant insulation

Figure 8.83 Sound and thermal insulation

Joist and stud coverings

Joists are decked with a suitable flooring material and the underside of the joists is usually covered with plasterboard. Stud partitions are also usually covered with plasterboard, with the occasional exception. For example, areas such as shower rooms are likely to be tiled, so a sturdier board such as WBP plywood is used.

Joist and stud coverings have come a long way in the past few years and the traditional method of lath and plaster is now only really used in listed buildings.

Method 1

The plasterboard is fixed to the stud or joist with the back face of the plasterboard showing. The plasterer covers the whole wall or ceiling with a thin skim of plaster, leaving a smooth finish.

Method 2

The plasterboard is fixed with the front face showing, and the plasterer uses a special tape to cover any joints, then a ready mixed filler is applied over the tape and is used to fill in the nail and screw holes. Once the taped area is dry, the plasterer then gives the area a light sand to even it out.

Plasterboard comes with a choice of two different edges, and the right edge must be used:

- square edge – for use with Method 1; the whole wall/ceiling is plastered
- tapered edge – for use with Method 2, so that the plasterer can fix the jointing tape.

As well as having the correct edging, the right type of plasterboard must be used.

Various types of plasterboard are available so a colour coding system is used, although this may vary from manufacturer to manufacturer:

- fire-resistant plasterboard (usually red or pink) – used to give fire-resistance between rooms such as party walls

Figure 8.84 Lath and plaster

Figure 8.85 Plastered wall

Figure 8.86 Taped wall

- moisture-resistant plasterboard (usually green) – used in areas where it will be subjected to moisture such as bathrooms and kitchens
- vapour-resistant plasterboard – with a thin layer of metal foil attached to the back face to act as a vapour barrier, usually used on outside walls or the underside of a flat roof
- sound-resistant plasterboard – used to reduce the transfer of noise between rooms and can also help acoustics, used in places such as cinemas
- thermal check board – comes with a foam or polystyrene backing, so used to prevent heat loss and give good thermal insulation
- tough plasterboard – stronger than average plasterboard, it will stand up to more impact, used in public places such as schools and hospitals.

Sometimes plasterboard is doubled up – the walls are 'double sheeted' – to give fire-resistance and sound insulation without using special plasterboard.

Plasterboard comes in a variety of sizes, but the most standard size is 2400 × 1200 × 12.5 mm.

Cutting plasterboard

Plasterboard can be cut in two main ways: with a plasterboard saw or with a craft knife. The plasterboard saw can be used when accuracy is needed, but most jobs will be touched up with a bit of plaster, so accuracy is not vital. Using a craft knife is certainly quicker.

This is how to cut plasterboard with a craft knife:

Step 1 Mark the line to be cut with a pencil.

Step 2 Score evenly along the line with the craft knife ensuring hands, legs, etc. are out of the way of the blade.

Step 3 Turn the board round and give the area behind the cut a slap, to split the board in two.

Step 4 Run the knife along the back of the cut to separate the two pieces of board.

Step 5 Trim the cut with a slight back bevel to give a neater finish.

Figure 8.87 Step 1 Mark the line to be cut

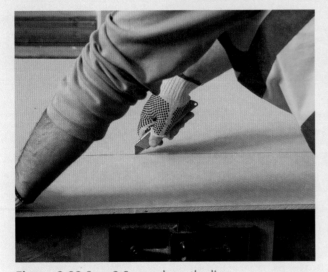

Figure 8.88 Step 2 Score along the line

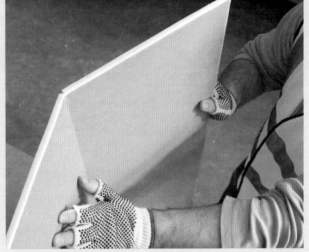

Figure 8.89 Step 3 Split the board in two

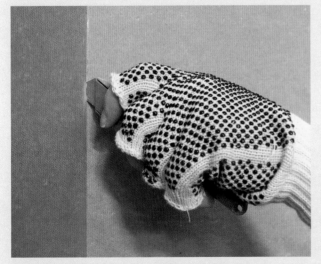

Figure 8.90 Step 4 Separate the pieces

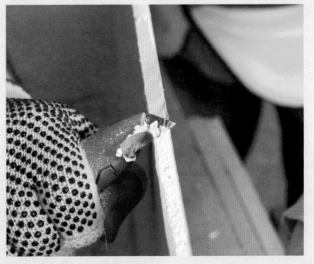

Figure 8.91 Step 5 Trim the cut

Fixing plasterboard

Plasterboard can be fixed by three different methods, which are:

- nail – plasterboard can be fixed using clout or special plasterboard nails. When nailing, make sure the nail is driven below the surface with the hammer, leaving a dimple or hollow that can be filled by plaster later. The nails used must be galvanised. If they are not galvanised, the nails will rust and 'bleed' when the plaster is put on to fill the hollows – the rust from the nails will stain the plaster and can even show through several coats of paint

- screw – screws can be used but these must also be galvanised to prevent bleeding

- dot and dab – this is used where the plasterboard is fitted to a pre-plastered or flat surface such as blockwork. With dot and dab, there is no need for a stud or frame to fix the plasterboard, thus increasing the room size (it is often used in stairwells where space is limited). Dot and dab involves mixing up plaster and dabbing it onto the back of the board, pushing the board directly onto the wall.

> **Remember**
>
> Dot and dab is a skilled technique, as with this method getting the boards plumb and flat can be difficult

K4: Know how to assemble, erect and fix straight flights of stairs including handrails

Stairs are a means of providing access from one floor to another. They are made up of a number of steps and each continuous set of steps running in one direction is called a flight. Hence, a staircase may have several flights. They are often the cause of accidents in buildings and are governed very strictly by Building Regulations.

Stairs in this section will be looked at under the headings of:

- terminology
- building regulations
- setting and marking out
- how to cut and construct stairs
- how to install them.

Terminology

There are a number of terms and definitions associated with construction of stairs, which you will need to know and understand. The main ones are listed below and shown in Figure 8.92.

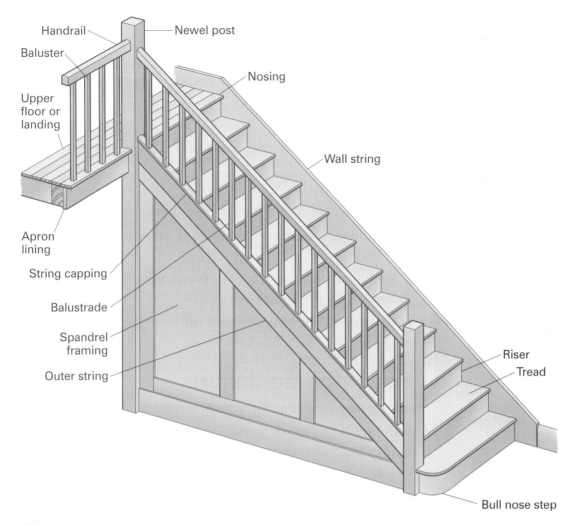

Handrail

Newel post

Baluster

Upper floor or landing

Nosing

Wall string

Apron lining

String capping

Balustrade

Spandrel framing

Outer string

Riser

Tread

Bull nose step

Figure 8.92 Main stair parts

- **landings** – between floor levels to break up the overall length of a flight and can be used to change the direction of a flight of stairs
- **string** – main board to which treads and risers are fixed, including wall string, outer string, close string and cut string
- **tread** – flat, horizontal part of the step
- **riser** – vertical part of the step
- **rise** – the height of the step (the measurement from the top of one step to the top of another)
- **step** – combination of one tread and one riser
- **going** – horizontal distance measured from the nosing of one step to the nosing of the step directly above or below
- **newel post** – heavy vertical member at each end of the stair to which the handrail is fixed

Figure 8.93 Straight flight stairs

- **balustrade** – unit comprising handrail, newels and the infill between it and the string, which provides a barrier for the open side of the stair
- **balusters** – vertical members forming the infill between the string and the handrails
- **bull nose step** – quarter-rounded step at the bottom of a stair
- **stairwell** – opening formed in the floor layout to accommodate a stair
- **nosing** – front edge of a tread, or loose narrow top tread, which sits on the trimmer joist at the top of the stairs
- **capping** – fixed on the top edge of strings to take the fixing of a balustrade
- **cap** – shaped top of a newel post, which can be fixed on or turned on the solid newel
- **spandrel framing** – where the triangular area is formed under the stairs, which can be framed to form a cupboard.

Building Regulations

The construction and design of a stair is controlled under Building Regulations 2000: Part K. These specify the requirements for different types of stairs depending on the type of building and what it is to be used for.

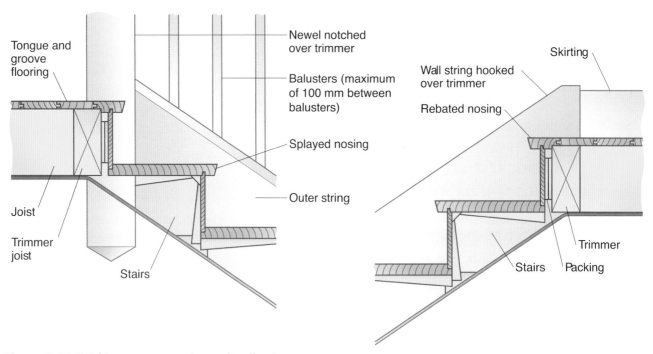

Figure 8.94 Finishings to outer string and wall string

Tables 8.3 to 8.5 and Figures 8.93–8.95 detail some of the requirements in the regulations.

Description of stair (to meet)	Max rise (mm)	Min going (mm)	Range (to meet pitch limitation) (mm)
Private stair	220	220	155–220 rise with 245–260 going or 165–200 rise with 220–350 going
Common stair	190	350	155–190 rise with 240–320 going
Stairway in institutional building (except stairs used only by staff)	180	280	
Stairway in assembly area (except areas under 100 m²)	180	250	
Any other stairway	190	250	

Table 8.3 Regulations for rise and going

Description of stair	Minimum balustrade height (mm)	
	Flight	Landing
Private stair	840	900
Common stairway	900	1000
Other stairway	900	1100

Table 8.4 Regulations for minimum balustrade heights

Description of stair	Minimum width (mm)
Private stair giving access to one room only (except kitchen and living room)	600
Other private stair	800
Common stair	900
Stairway in institutional building (except stairs used only by staff)	1000
Stairway in assembly area (except areas under 100 m²)	1000
Other stairway serving an area that can be used by more than 50 people	1000
Any other stairway	800

Table 8.5 Regulations for width of stairs

Functional skills

Using the information contained in these tables will give you the opportunity to practice FM 1.2.1b relating to interpreting information from sources such as diagrams, tables, charts and graphs.

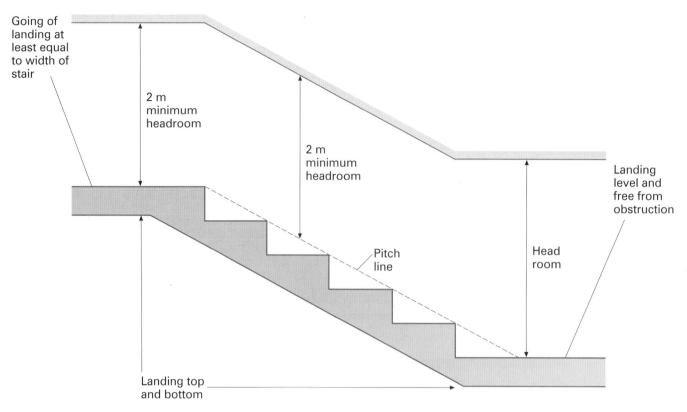

Figure 8.95 Regulations for minimum headroom height and landing requirements

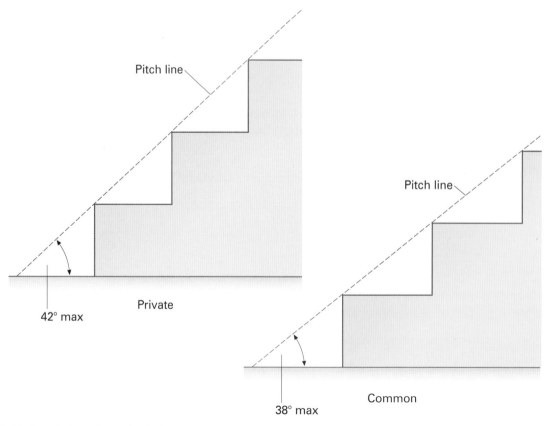

Figure 8.96 Regulations for stair pitch

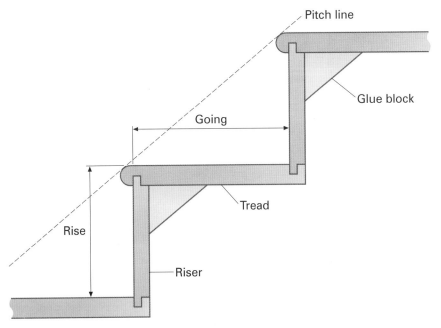

Figure 8.97 Rise and going

Figure 8.94 shows how the pitch or steepness is limited to a maximum of 42° for a private stair and a maximum of 38° for a common stair (i.e. where a stairway is used by more than one dwelling).

The steps for a straight flight of stairs should all have the same rise and going. See Figure 8.95. There are limits to these dimensions for different stairs. In all situations twice the rise plus the going should work out between 550 and 700 mm (2R+G).

Installation of stairs

Installing a staircase on a construction site is a major part in the construction of the building. The same basic procedures are followed, though there will be differences depending on what type of staircase is to be installed, as well as variations due to the site.

A staircase is a large and fairly expensive item and should be handled with care when delivered to site and offloaded. They are normally delivered in completed form as far as possible, but components such as the handrail, balusters and newel posts will have to be fitted to suit the site requirements.

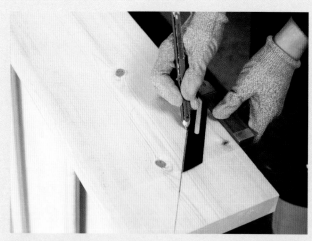

Figure 8.98 Step 1 Fix the wall string

Figure 8.99 Step 2 Cut the seat cut

Figure 8.100 Step 3 Check level

Figure 8.101 Step 4 Cut away the top tread

Figure 8.102 Step 5 Mortise the outer string

Figure 8.103 Step 6 Fix the newel posts

Figure 8.104 Step 7 Fix the wall string to the wall

Step 1 Fix the wall string, cutting it off at floor level to suit the skirting height.

Step 2 Use a hand saw to cut the seat cut on the string at the foot of the stairs.

Step 3 The staircase is now level on the floor.

Step 4 Steps 1 and 2 can be repeated at the top with the underside cut out to sit on to the floor trimmer and the top tread is cut away so that it sits on the trimmer.

Step 5 Mortise the outer string into the newel posts at each end.

Step 6 Fix newel posts in place. The bottom newel can be held in position using various methods and this will depend on the composition of the floor. For rigidity and maximum strength the top newel should be notched over the trimmer joist and screwed or bolted to it.

Step 7 Fix the wall string to the wall in approximately four places below the steps, usually with 75 mm screws and plugs. Fixing should be below the steps, unless access is difficult in which case the resulting hole should be plugged. The balustrades and hand rail can be fitted once the stairs are secure.

Step 8 Once the staircase has been fitted, it should be protected to prevent damage. Strips of hardboard should be pinned to the top of each tread with a lath to ensure the nosing is protected. Use the same method to protect the newel posts.

Figure 8.105 Step 8 Protect from damage

Working life

Finishing off

Scott and Nick are working in a large house plasterboarding the stud partition walls. When they come to the last wall, they find a note left by the plumber saying 'Don't plasterboard'. They are on price work and Nick says that they should just finish the work and get paid. Scott says that they should ask the plumber first.

- Who is right?
- What could the consequences be?

Functional skills

Working with other trades is an important part of working on site. Dealing with problems like this will give you the opportunity to practice the *Interpreting* elements of functional skills, e.g. FM 1.3.1 Judge whether findings answer the original problem, FM 1.3.2 Communicate solutions to answer practical problems.

FAQ

I have laid chipboard flooring and the floor is squeaking. What causes this and how can I stop it?

The squeaking is caused by the floorboards rubbing against the nails – something that happens after a while as the nails eventually work themselves loose. The best way to prevent this is to put a few screws into the floor to prevent movement. Be cautious: there may be wires or pipes under the floor.

My maths isn't very good so I often struggle with the geometry and trigonometry skills needed, especially for roofing. What can I do?

Your training will include some basic numeracy skills lessons, and if you struggle with maths you should take advantage of them. Be patient and pay attention to any help and advice you are given regarding these sometimes complicated calculations. You will probably find that the more practical experience you get, the better your maths skills will become.

Check it out

1   Draw a sketch of a window, labelling the component parts.
2   What is a mortar key used for in a window frame?
3   Explain why it is important to assemble windows flat.
4   Write a method statement explaining how you can check whether a window frame is square while you assemble it.
5   Name two ways of fixing a window board into position.
6   How do you level a window board?
7   State and describe three materials used as window boards.
8   Explain what types of weight are used in a box sash window.
9   Name three types of bay window.
10 List five components you will find on a flight of stairs.
11 What is the minimum headroom allowed as stated in the *Building Regulations* covering stairs?
12 Name the vertical member used to provide support to a handrail and infill on an open balustrade.
13 After stairs have been fitted, what three measures can be taken to prevent damage during building work?
14 Name three ways of fixing grounds.
15 What PPE should be worn when using masonry nails?

Getting ready for assessment

The information contained in this unit, as well as continued practical assignments that you will carry out in your college, or training centre, will help you with preparing for both your end of unit test and the diploma multiple-choice test. It will also aid you in preparing for the work that is required for the synoptic practical assignments.

The information in this unit will aid you in learning how to identify and calculate the materials and equipment required for first fixing.

You will need to be familiar with:

- preparing materials for carrying out a range of first fixing tasks
- fixing frames and linings
- fitting and fixing floor coverings and flat roof decking
- erecting timber stud partitions
- assembling, erecting and fixing straight flights of stairs.

This unit will have made you familiar with these common first fixing tasks. For learning outcome two, you have seen the types and sizes of timber that can be used for floor coverings and flat roof decking. You will need to use your knowledge of these types of timber to select the correct one for any particular fixing job you may be working on. You will also need to select the correct hand tools and equipment to work with a particular material in a particular situation.

This unit has shown the types and methods of fixing joist coverings and forming openings to services under floors. You will now need to use this knowledge in your practical tasks to install floor coverings and flat roof coverings accurately, and to working drawings and specifications. You will also have to identify the best methods of leaving openings for services, as well as being able to identify any problems with the service openings.

Before you start work on the synoptic practical test it is important that you have had sufficient practice and that you feel that you are capable of passing. It is best to have a plan of action and a work method that will help you. You will also need a copy of the required standards, any associated drawings and sufficient tools and materials. It is also wise to check your work at regular intervals. This will help you to be sure that you are working correctly and help you to avoid problems developing as you work.

Your speed at carrying out these tasks will also help you to prepare for the time limit that the synoptic practical task has. But remember, don't try to rush the job as speed will come with practice and it is important that you get the quality of workmanship right.

Always make sure that you are working safely throughout the test. Make sure you are working to all the safety requirements given throughout the test and wear all appropriate personal protective equipment. When using tools, make sure you are using them correctly and safely.

Good luck!

CHECK YOUR KNOWLEDGE

1   The purpose of a window is:
    a   to allow natural light into a room.
    b   to conserve heat.
    c   to protect from weather.
    d   all of the above.

2   When assembling a window frame, you can check that it is square by:
    a   measuring the length and width.
    b   using the 3:4:5 method.
    c   using a squaring rod.
    d   using a spirit level.

3   A window that projects from the front of a building is called a:
    a   bay window.
    b   stormproof window.
    c   casement window.
    d   box sash window.

4   The vertical part of a step on a staircase is called a:
    a   tread.
    b   riser.
    c   newel.
    d   string.

5   The combination of a tread and a riser is called a:
    a   stair.
    b   string.
    c   step.
    d   stairwell.

6   On a staircase, a string is:
    a   an opening formed in the floor to accept a stair.
    b   something the handrail is fixed to.
    c   the height of a step.
    d   the board that the treads and risers are fitted to.

7   The maximum pitch for a private stair is:
    a   38°.
    b   40°.
    c   42°.
    d   44°.

8   The top of a staircase must be fixed to a:
    a   trimmer joist.
    b   trimmed joist.
    c   trimming joist.
    d   bridging joist.

9   The timber part of a stud partition that stiffens the structure and allows a fixing for heavy components is called:
    a   nogging.
    b   nugging.
    c   nigging.
    d   nagging.

10 Extra studs are placed at corners because:
    a   they are stronger.
    b   they offer a fixing for cladding.
    c   they are weaker.
    d   they allow for doorways.

11 Partitions can be made more fire-resistant by:
    a   using insulation.
    b   double boarding.
    c   using larger studs.
    d   using more studs.

12 A frame that runs from floor to floor is called a:
    a   doorframe.
    b   fan light frame.
    c   storey frame.
    d   door lining.

13 The joint used in the construction of a door lining is a:
    a   mortise and tenon joint.
    b   butt joint.
    c   housing joint.
    d   tongued housing joint.

14 The purpose of grounds is:
    a   to provide a flat surface to allow cladding.
    b   to allow easy fixing for skirting boards.
    c   to provide a line for the plasterer to work to.
    d   all of the above.

Know how to carry out second fixing

Once plasterwork has been completed, you will be in a position to carry out second fixing. Second fixing is the name given to all joinery work undertaken after plasterwork has been finished, and includes side-hung doors, units, and ironmongery. In this unit you will learn the skills needed to confidently complete a second fixing.

This unit also supports NVQ Unit VR10 Install Second Fixing Components.

This unit contains material that supports TAP unit 3 Install Second Fixing Components. It also contains material that supports the delivery of the five generic units.

This unit will cover the following learning outcomes:

- How to install side hung doors and ironmongery
- How to install mouldings
- How to install service encasements and cladding
- How to install wall, floor units and fitments.

Functional skills

When reading and understanding the text in this unit, you are practicing several functional skills.

FE 1.2.1 – Identifying how the main points and ideas are organised in different texts.

FE 1.2.2 – Understanding different texts in detail.

FE 1.2.3 – Read different texts and take appropriate action, e.g. respond to advice/ instructions.

If there are any words or phrases you do not understand, use a dictionary, look them up use the internet or discuss with your tutor. FM 1.2.1b relates to interpreting information from sources such as diagrams, tables, charts and graphs

Key term

Stile – the longest vertical timber in the frame of a door, etc.

Did you know?

A material that swells when subjected to heat or damp is called intumescent

K1. Know how to install side hung doors and ironmongery

Internal doors

This section has been designed to provide the knowledge and understanding to select and hang internal doors. It will cover:

- the purpose of internal doors
- the types of internal door
- choosing a door
- fitting a door.

Purpose of internal doors

The purpose of an internal door is to provide a means of privacy, a thermal, sound or security barrier, a means of access and egress (i.e. a way out) and, in some instances, a fire-resisting barrier.

Types of internal door

Framed doors

Framed doors are doors made from hardwood or softwood and are constructed using mortise and tenon joints (described in Unit 2008), or dowelled joints. The frame is rebated or grooved, into which a board panel can be fitted. Alternatively glass could be used and held in place with beading. Figure 9.1 shows an exploded view of a framed door, including the types of joint used.

Figure 9.2 also shows types of sections used for door panels.

Flush doors

Flush doors are lightweight, cheap and simply made. Most consist of a softwood frame, which is stapled together and houses a hollow-core material (usually cardboard honeycomb that has the appearance of egg crates). They are then faced with hardboard or plywood. See Figure 9.3.

Flush doors often come with fitting instructions and these indicate which stile contains a lock block (an extra block of wood included in the door frame to take the door lock). Failing this, a symbol is normally printed on the head or bottom rail that indicates where the lock block can be found.

Fire-resisting doors

Fire-resisting doors have a core of solid, fire-retarding material and are frequently flush panelled. Their ratings are designated by

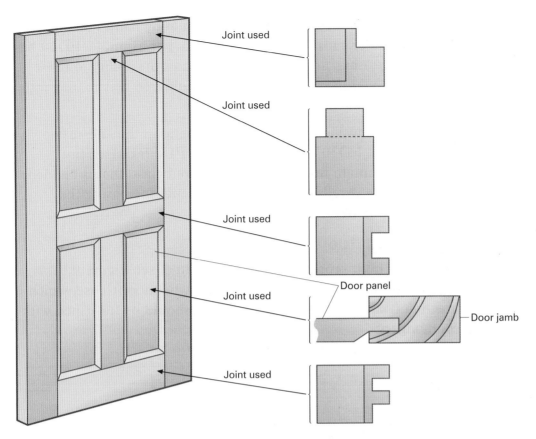

Joint used

Joint used

Joint used

Door panel

Joint used

Door jamb

Joint used

Figure 9.1 Framed door exploded view

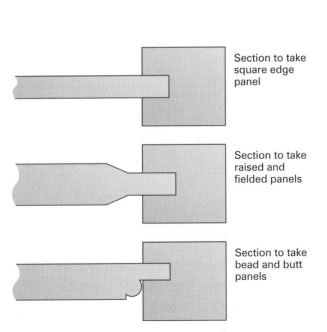

Section to take square edge panel

Section to take raised and fielded panels

Section to take bead and butt panels

Figure 9.2 Sections used for door panels

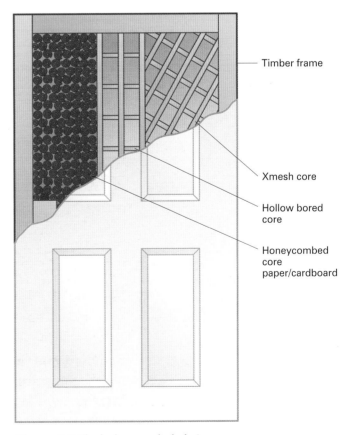

Timber frame

Xmesh core

Hollow bored core

Honeycombed core paper/cardboard

Figure 9.3 Flush door exploded view

Figure 9.4 Fire door plugs – 30- and 60-minute (note that colours used differ between manufacturers)

their performance (resistance to penetration by flames or smoke through splits or gaps) with the prefix FD. For example, FD30 has a 30-minute rating, and an S suffix indicates the ability to resist smoke. The manufacturers also insert a coloured plastic plug into the door frame to indicate the rating.

Fire doors are available in standard sizes and in thicknesses of 44 mm for the FD30 and 54 mm for the FD60. Glazed fire doors must contain fire-rated plain or wired glass, that is bedded in material that swells up when heated, hence increasing the resistance to smoke.

Choosing a door

On a newly built property the door schedule (see Figure 9.5) and plans of the building will determine what door to use, which direction it shall operate and also how it should be furnished with ironmongery.

Functional skills

Using schedules allows you to practice FM 1.2.1b relating to interpreting information from sources such as diagrams, tables, charts and graphs.

Description	D1	D2	D2	D2	D2	D2	D2	D2	D2	D10	NOTES
Type											
External glazed A1					●						
External panel A2	●										
Internal flush B1									●		
Internal flush B2		●				●	●	●		●	
Internal glazed B3			●	●							
Size											
813 mm x 2032 mm x 44 mm	●				●						
762 mm x 1981 x 35 mm		●	●	●		●	●	●		●	
610 mm x 1981 x 35 mm									●		
Material											BBS DESIGN
Hardwood	●										
Softwood			●	●	●						
Plywood/Polished		●									JOB TITLE
Plywood/Painted						●	●	●	●	●	
Infill											DRAWING TITLE Door schedule/doors
6 mm tempered safety glass											JOB NO. DRAWING NO.
Clear			●	●	●						
Obscured	●										SCALE DATE DRAWN CHECKED

Figure 9.5 A door schedule

Modern buildings tend to have standardised joinery throughout. Standard door heights are 2 m, 2.03 m and occasionally 2.17 m; widths range from 600 mm to 900 mm, usually in 75 mm increments. Thicknesses vary from 35 mm to 44 mm. Standard single door sizes that are available by type of door from a typical manufacturer are shown in Table 9.1. They include several that match the old imperial sizes (quoted in millimetres), as these are still required to replace existing doors. There are additional standard sizes available for double leaf and single and double garage doors.

Table 9.1 Standard single door sizes (mm)

Height	1981	1981	1981	1981	2000	2032	2040	2040	2040
Width	610	686	762	838	807	813	626	726	826
External	*	*	*	*	*	*		*	*
Internal	*	*	*	*			*	*	*

Older properties often have relatively large doors to main rooms and may be of a non-standard size. In this case, the door will have to be made to measure, the nearest available size or, larger, purchased and trimmed to fit by removing an equal amount from each stile to keep the frames symmetrical.

Fitting a door

Whenever possible doors should be stored flat for a few days prior to fitting in the location where they are to be hung. This is sometimes referred to as 'storing out of twist'. This gives the timber time to acclimatise to its new surroundings and any final shrinkage or reduction in moisture content can take place (which sometimes happens with modern central heating systems).

The following actions should be completed before fitting a door:

- check the building plans and the door schedule to determine which type of door is to be used and in which direction it should operate. Failing this, ask the supervisor or client
- check that the frame is square and aligned properly
- measure the new door to make sure it will fit within the frame and check that it is not twisted
- cut off horns, if fitted
- find where the lock block is located. The other stile is the hanging stile to which hinges will be attached.

Did you know?

Doors should normally conceal the largest area of the room when open to allow maximum privacy to those inside

Fitting is carried out as follows:

Did you know?

Because the hinge knuckle goes on the outside edge of the door, the arc followed by the inside edge is different; so a door requires a bevelled lead-in edge to close properly

Step 1 There must be a gap of 2 mm between the head of the door and the head of the frame and at least 6 mm clearance between the floor covering and the bottom of the door. If the door is too tall, always cut excess off the bottom rail.

Step 2 There must be a gap of 2 mm between both door styles and the frame. If necessary, plane evenly from both sides of the door. The door should now fit in the frame. With the

Figure 9.6 Step 1 Make sure there is a gap between the door and the frame

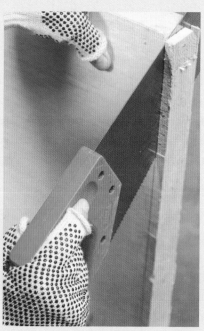

Figure 9.7 Step 2 Cut door to width and shape styles

Figure 9.8 Step 3 Bevel the leading edge

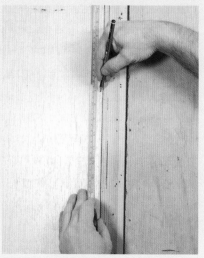

Figure 9.9 Step 4 Mark hinges

hanging stile against the frame, wedge the door up so there is a 2 mm gap at the top and mark, if necessary, to shape the hanging style to match the frame. Plane off as necessary.

Step 3 If necessary, mark and shape the opposite stile (containing the lock block), then slightly bevel the edge that will lead into the frame so that it will not catch when closing.

Step 4 With the door fitting snugly in the frame and wedged up to give 2 mm clearance at the top, mark where the hinges are to go, on both door and frame at the same time. The top hinge normally sits 150 mm from the top of the door and the bottom hinge 225 mm from the bottom of the door.

Step 5 Remove the door from the frame and accurately mark where the hinges are to go, with the aid of a square and marking gauge. Do the same on the frame and then chop out recesses on both door and frame.

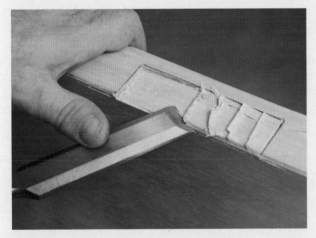

Figure 9.10 Step 5 Mark and cut hinge recesses

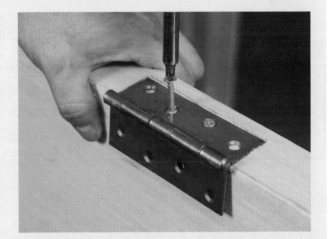

Figure 9.11 Step 6 Fit hinges

Step 6 Fit the hinges into the door recesses, putting in all the screws. Take the door to the frame and place the hinges into the frame recesses, securing the hinges using only one screw. Check the door swings without binding. Alter or adjust until there is an equal gap on both sides and the correct clearance at the top and no resistance is encountered when closing the door. Put in all remaining screws on the stile hinges.

Step 7 Fit all remaining ironmongery, or furniture, as instructed by the schedule or client. (Ironmongery is covered in the next section.)

Figure 9.12 Step 7 Fit all remaining ironmongery

External doors

This section aims to provide the knowledge and understanding needed to select, hang and fix the required ironmongery to an external door.

Types of external door

External doors are always solid or framed – a flush or hollow door will not provide the required strength or security.

In some cases external doors are made from uPVC. uPVC doors require little maintenance and the locking system normally locks the door at three or four different locations, making it more secure. High-quality external doors are usually made from hard-wearing hardwoods such as mahogany or oak, but can be made from softwoods such as pine.

External doors come in the same dimensions as internal doors, except external doors are thicker (44 mm rather than 40 mm). On older properties external doors often have to be specially made as they tend not to be of standard size.

There are four main types of external door:

- framed, ledged and braced
- panelled doors
- half- and full-glazed doors
- stable doors.

> **Did you know?**
>
> A framed, ledged and braced door comes with the bracing unattached. The bracing can then be attached to suit the side on which the door is hung

Framed, ledged and braced doors

Framed, ledged and braced (FLB) doors consist of an outer frame clad on one side with tongued and grooved boarding, with a bracing on the back to support the door's weight.

An FLB door is usually used for gates and garages and sometimes for back doors. When hanging an FLB door it is vital that the bracing is fitted in the correct way; if not, the door will start to sag and will not operate properly.

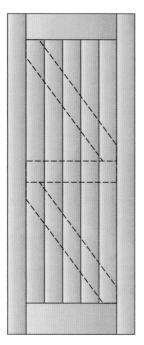

Front elevation

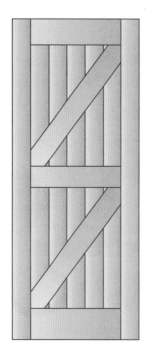

Back elevation

Figure 9.13 Front and back of a framed, ledged and braced door

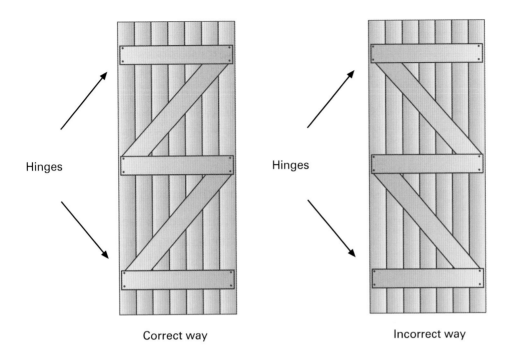

Hinges

Hinges

Correct way

Incorrect way

Figure 9.14 Bracing fitted correctly and incorrectly

Panelled doors

Panelled doors consist of a frame made up from stiles, rails, muntins and panels. Some panel doors are solid, but most front and back doors have a glazed section at the top to allow natural light into the room.

Half- and full-glazed doors

Half-glazed doors are panelled doors with the top half of the door glazed. These doors usually have diminished rails to give a larger glass area.

Full-glazed doors come either fully glazed or with glazed top and bottom panels, separated by a middle rail.

Full-glazed doors are used mainly for French doors or back doors where there is no need for a letterbox.

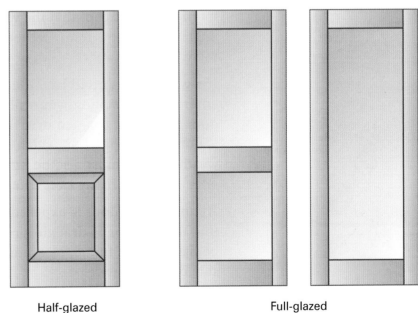

Half-glazed

Full-glazed

Figure 9.15 Half-glazed door and two types of full-glazed door

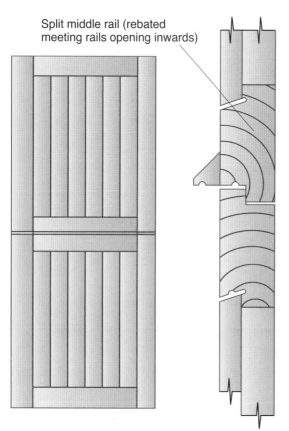

Split middle rail (rebated meeting rails opening inwards)

Figure 9.16 Stable door with section showing the rebate on the middle rails

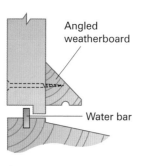

Angled weatherboard

Water bar

External

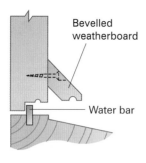

Bevelled weatherboard

Water bar

Internal

Figure 9.17 Internally and externally opening doors

Stable doors

As their name implies, stable doors are modelled on the doors for horses' stables and they are now most commonly used in country or farm properties. A stable door consists of two doors hung on the same frame, with the top part opening independently of the bottom. The make-up of a stable door is similar to the FLB door, but the middle rail will be split and rebated as shown in Figure 9.16.

To hang a stable door, first secure the two leaves together with temporary fixings, then hang just like any other door, remembering that four hinges are used instead of three.

Hanging an external door

External doors usually open inwards, into the building. Where a building opens directly onto a street, this prevents the door knocking into unsuspecting passers-by, but, even where there is a front garden, having a door that opens outwards is not good practice as callers will need to move out of the way to allow the door to be opened. Externally opening front doors are usually only used where there is limited space, or where there is another door nearby which affects the front door's usage.

Hanging an exterior door is largely the same as hanging an interior door, except that the weight of the door requires three hinges sited into a frame rather than a lining. Because of this, there is usually a threshold, or sill, at the bottom of the doorframe.

If a water bar is fitted into the threshold to prevent water entering the dwelling, you will need to rebate the bottom of the door to allow it to open over the water bar. The way the door is hung will depend on which side of the door is rebated. A weatherboard must also be fitted to the bottom of an external door, to stop driving rain entering the premises.

As an exterior door is exposed to the elements it is important that the opening is draught proofed: a draught-proofing strip can be fitted to the frame, or draught proofing can be fitted to the side of the door.

Ironmongery for an external door

An external door requires more ironmongery than an internal door, and may need:

- hinges
- letter plate
- mortise lock/latch
- mortise dead lock
- cylinder night latch
- spy hole
- security chains.

Hinges

External door hinges are usually butt hinges, though FLB doors often use T hinges. Three 4-inch butt hinges are usually sufficient, though it is advisable to use security hinges (hinges with a small steel rod fixed to one leaf, with a hole on the other leaf) to prevent the door being forced at the hinge side.

Letter plate

The position of a letter plate depends on the type of door. Letter plates are usually fitted into the middle rail, but could be fitted in the bottom rail (for full-glazed doors) or even the stile (with the letter plate fitted vertically).

A letter plate can be fitted either by drilling a series of holes and cutting out the shape with a jigsaw, or by using a router with a guide.

Once the opening has been formed, the letter plate can be screwed into place.

Mortise lock / latch

Mortise locks are locks housed into the stile and are fitted as follows:

> **Did you know?**
>
> Draughts can occur in the letter plate area, so it is good practice to have a draught-proofing strip fitted

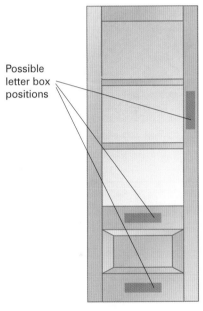

Possible letter box positions

Figure 9.18 Different options for fitting a letter plate

Step 1 Mark out the position for the lock (usually 900 mm from the floor to the centre of the spindle) and mark the width and thickness of the lock on the stile.

Step 2 Using the correct sized auger/flat bit, drill a series of holes to the correct depth.

Figure 9.19 Step 1 Mark out the position

Figure 9.20 Step 2 Drill a series of holes

Step 3 Using a sharp chisel remove the excess timber, leaving a neat opening.

Step 4 Mark around the faceplate and with a sharp chisel, remove the timber so that the faceplate sits flush with the stile.

Step 5 Mark where the spindle and keyhole are, then drill out to allow them both to be fitted and operate properly.

Step 6 Fix the lock and handles in place.

Step 7 Mark the position of the striking plate on the doorframe, then house it into the frame.

Step 8 Check that the lock operates freely.

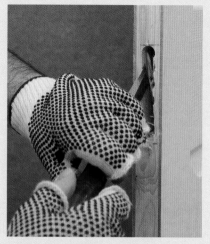

Figure 9.21 Step 3 Remove excess timber

Figure 9.22 Step 4 Mark around the faceplate

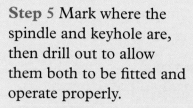

Figure 9.23 Step 5 Drill out spindle and key hole

Figure 9.24 Step 6 Fix the lock and handles

Figure 9.25 Step 7 House the striking plate

A mortise lock on its own does not usually provide sufficient security for an exterior door, so most doors will also have one of the following:

- **mortise deadlock** – fitted like a mortise lock except that it has no latch or handles, so an escutcheon is used to cover the keyhole opening. It is usually fitted three-quarters of the way up the door

- **cylinder night latch** – preferred to a mortise deadlock as it does not weaken the door or frame as much. It is also usually fitted three-quarters of the way up the door. The manufacturer will provide fitting instructions with the lock.

Spy hole

A spy hole is usually fitted to a door that is solid or has no glass and is used as a security measure so that the occupier can see who is at the door without having to open the door.

To fit a spy hole, simply drill a hole of the correct size, unscrew the two pieces, place the outer part in the hole, then re-screw the inner part to the outer part.

Security chain

A security chain allows the door to be opened a little without allowing the person outside into the dwelling. The chain slides into a receiver, then as the door opens the chain tightens, stopping the door from opening too far.

Double doors

Two doors are fitted within one single larger frame/lining, with 'meeting stiles', that is two stiles that meet in the middle, rebated so that one fits over the other.

Double doors are usually used where a number of people will be walking in the same direction, while double swing doors are used where people will be walking in different directions. Double doors allow traffic to flow without a 'bottleneck' effect and let large items such as trolleys pass from one room to another. Double doors are used in places such as large offices, or public buildings such as schools and hospitals.

Hanging double doors is the same as for any other door, though extra care must be taken to ensure that the stiles meet evenly, and that there is a suitable gap around the doors.

Did you know?

Spy holes work by having a small curved lens, which creates a wider field of vision than normal. This means that, even though the hole is small, you can see a full-height image of the person standing outside the door. The lens bends the image so that it appears curved, like the reflection in a festive bauble

Figure 9.26 Basic security ironmongery

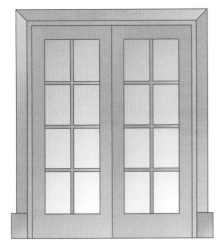

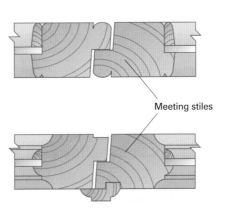

Meeting stiles

Figure 9.27 Meeting stiles

Figure 9.28 Parliament hinge

Figure 9.29 Rebated mortise lock

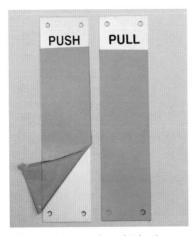

Figure 9.30 Push or kick plates

Key terms

Helical – in a spiral shape

Helical spring hinge – a hinge with three leaves, which allows a door to be opened through 180 degrees

Extra ironmongery is required on double doors as follows:

- parliament hinges – these project from the face of the door and allow the door to open 180 degrees
- rebated mortise lock – similar to a standard mortise lock, but the lock and striking plate are rebated to allow for the rebates in the meeting stiles
- push/kick plates – fixed to the meeting stiles and the bottom rails to stop the doors getting damaged
- door pull handles – fixed to the meeting stiles on the opposite side from the push plates, these allow the door to be opened
- barrel bolts – usually fixed to one of the doors at the top or the bottom to secure the door when not in use.

Figure 9.31 Pull handles

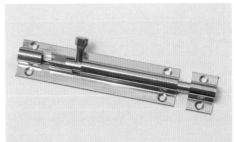

Figure 9.32 Barrel bolt

Most double doors also have door closers to ensure that they will close on their own, to prevent the spread of fire or draughts throughout the building.

There are four main types of door closer: concealed, overhead, floor springs and **helical spring hinges**. Floor springs and **helical** hinges will be looked at on page 184.

Concealed door closers

Concealed door closers work through a spring and chain mechanism housed into a tube.

Figure 9.33 Concealed door spring closer

To fit a concealed door closer, you must house the tube into the edge of the hinge side of the door, with the tension-retaining plate fitted into the frame.

Overhead door closers

As the name implies, overhead door closers are fitted to either the top of the door or the frame above. They work through either a spring or a hydraulic system fitted inside a casing, with an arm to pull the door closed.

There are different strengths of overhead door closer to choose from depending on the size and weight of the door. Table 9.2 below shows the strengths available.

Table 9.2 Strengths of overhead door closures

Power no.	Recommended door width	Door weight
1	750 mm	20 kg
2	850 mm	40 kg
3	950 mm	60 kg
4	1100 mm	80 kg
5	1250 mm	100 kg
6	1400 mm	120 kg
7	1600 mm	160 kg

> **Remember**
>
> When choosing an overhead door closer, you must take into account any air pressure from the wind. If the pressure is strong, you may require a more powerful closer

There are a number of different ways to fit an overhead door closer, depending on where the door is situated. The two main ways are:

- fitting the main body of the closer to the door, with the arm attached to the frame

- inverting the closer so that the main body is fitted to the frame and the arm is fitted to the door.

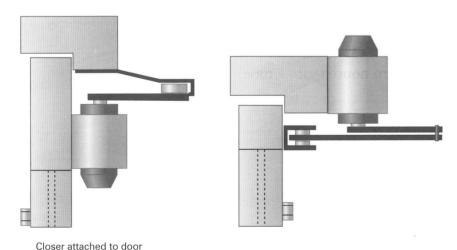

Closer attached to door

Figure 9.34 Closer fitted to door   **Figure 9.35** Closer fitted to frame

Figure 9.36 Helical or double action hinge

Door closers come with instructions and a template, to make fitting easier.

Double swing doors need to open both ways to accommodate traffic going in both directions. The main difference work-wise between standard double doors and double swing doors is in the ironmongery. Double swing doors need special hinges and must have some form of door closer fitted – it is usually best to combine these, in one of two ways:

- **helical spring hinges** – Helical spring hinges are three-leaf hinges with springs integrated into the barrels. The way the leaves are positioned allows the doors to open both ways. The hinges are fitted just like normal hinges and the tension on the springs are adjusted via a bar inserted into the hinge collar

- **floor springs** – Doors using floor springs are hung via the floor spring at the bottom and a pivot plate at the top. The floor spring is housed into the floor and the bottom of the door is recessed to accept the shoe attached to the floor spring.

The pivot plate at the top is attached to the frame and a socket is fixed to the top of the door.

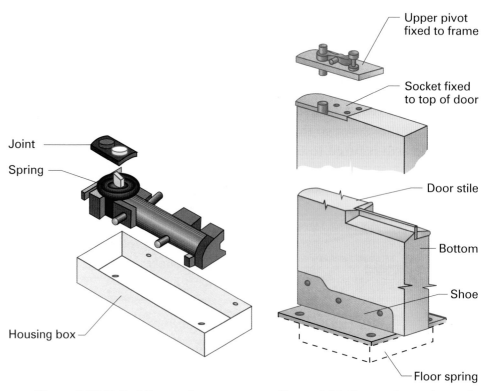

Figure 9.37 Helical floor springs

Figure 9.38 Floor spring and top of door fixing

General door ironmongery

This section has been designed to enable you to choose the correct fixtures and fittings for the work you are doing.

It offers just a small guide to what is available and the range of ironmongery is continually increasing. As a carpenter it will be your responsibility to choose and supply the correct items to your client. Therefore, you would be advised to look around your local DIY superstore to see what is available. Also look at completed work to see what other people have used.

Functions of ironmongery

Ironmongery is also referred to as hardware and relates to the components used to fix or decorate work.

To fix a door into a frame a carpenter would require a minimum of hinges, screws and some form of lock or latch, unless the client specified otherwise. All are ironmongery and can be classified as fixings that:

- allow movement
- provide security
- penetrate timber.

Fixings that allow movement

These are principally hinges and made of metal; the main ones are shown in Figures 9.39 to 9.43.

Standard cranked butt hinges are fitted so that the knuckle of the hinge protrudes beyond the face of the door and the edge of the door jamb. These hinges are available in sizes from 25 mm to 100 mm and are produced in brass or steel.

Cast butt hinges are produced from cast iron and are classified as heavy duty. They will rust if left untreated.

Rising butt hinges have a helical knuckle that allows the door to rise when opened, which assists the door in clearing carpets and uneven floors. The helical knuckle also allows the door to close under its own weight.

Loose pin butt hinges enable the door to be removed from the frame by removing the pin from the hinge knuckle.

Storm-proof hinges are used in the manufacture of casement timber windows and have a cranked appearance.

Figure 9.39 Standard cranked butt hinges

Figure 9.40 Cast butt hinges

Figure 9.41 Rising butt hinges

Figure 9.42 Loose pin butt hinges

Figure 9.43 Tee hinges

Key term

Gudgeons – tubes at the end of hinges to take the pin around which the hinge rotates

Figure 9.45 Bands and gudgeons

Figure 9.46 Mortise dead lock

Friction hinges are used on uPVC window systems and act as both a pivot and a stay.

Tee hinges are frequently used for hanging external doors on buildings like sheds. Light-duty versions are generally made from thin gauged steel and are either galvanised or painted in a black epoxy lacquer. Heavy duty versions are made of stronger steel with a steel or brass pin.

Figure 9.44 Friction hinge

Bands or straps of strong steel may be used instead of tee hinges for heavy-duty tasks, often doubled over and leaving a **gudgeon** to fit over a hinge pin welded to a plate on the frame, thus enabling the door or gate to be lifted off.

Fixings that provide security

Locks

Mortise dead locks have no latch and therefore they do not require standard handle furniture. Both the sash lock and dead lock can be obtained with hook bolts for use on sliding doors.

Mortise locks are available in 3-lever and 5-lever versions. 5-lever versions are stronger than their 3-lever counterparts and therefore more suitable for external door security. They also come in two depths: 75 mm for standard doors and 65 mm for narrow stile doors. Good-quality locks also have a reversible latch which means they can be fitted to left hand (LH) and right hand (RH) opening doors. Fitting instructions will be supplied with the locks and should be read thoroughly.

Figure 9.47 3- and 5-lever mortise locks

Figure 9.48 Cylinder lock

Cylinder locks consist of a body, a cylinder containing the key mechanism and a staple into which the latch engages.

Bolts and other fastenings

A range of bolts and other fastenings is available on the market. A selection of those most frequently used in second fixing is listed below.

A hasp and staple comprise a hinged strap (the hasp) that closes over a strong wire staple; a padlock is then put through the staple. It is normally used to secure garages and sheds. However, it is versatile, and is used for many applications.

Figure 9.49 Hasp and staple

Tower bolts consist of a long bar that is fixed to the door and is pushed to locate it in a keeper on the frame.

Pad bolts work on the same principle, but when the bolt is closed, the handle locates over a staple to which a padlock can be fixed.

Figure 9.50 Tower bolt

Bow handle bolts are much longer and more suited to heavy-duty applications, such as garage doors. The long bow handle means that the bolt can be operated even when the locking point is out of normal reach. They are made from square section metal and are spring loaded to prevent the bolt slipping out, especially under its own weight, if fitted vertically.

Monkey tail bolts are very similar to bow handle bolts, but have a ball on the end of the bolt, rather than a bow handle.

Figure 9.51 Pad bolt

Figure 9.52 Monkey tail bolt

Figure 9.53 Panic bar bolt

Panic bar bolts are available for single and double doors. They incorporate a bolt that runs the length of the door, passing through a latch that is fixed at approximately waist height. When the latch is pulled the bolts are in keepers positioned in the frame; when the latch is pushed the bolts disengage from the keepers. They are fitted on doors that are used as a means of escape, for example fire exits.

K2. Know how to install mouldings

This section will provide you with the knowledge and understanding required to enable you to select and fit internal mouldings.

Typical timber mouldings found within a building

Moulding refers to the pattern put on a length of timber. This is normally done by a machine such as a spindle moulder. However, specialist hand planes could be used to match existing patterns.

Common types of moulding that you will meet are:

- architrave
- skirting board
- plinth block
- picture rail
- dado rail
- cornice.

Where you will find these within a room is shown on Figure 9.54.

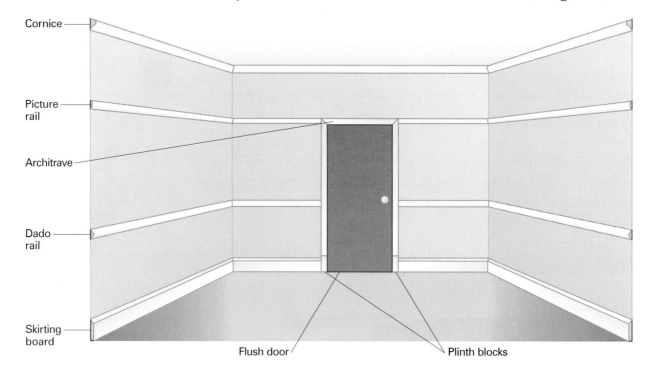

Figure 9.54 Types of moulding in a room

Mouldings are usually produced using premium grade timber, or medium density fibreboard (MDF), and should only be fixed at the 'second fixing' stage, once a building is weathertight and plastered.

Architrave

Architraves provide a decorative finish around internal openings, especially doors, and cover the joint between frame and wall finish. They are available in lengths ranging from 2.1 m to 5.1 m, increasing in 300 mm multiples; or as sets consisting of 2 × 2.1 m legs (the sides of the opening) and 1 × 900 mm head (the top of the opening). Architraves are between 50 mm and 75 mm wide and are usually 19 mm to 25 mm thick before they are planed. There is a range of commercially available architrave mouldings.

Figure 9.55 Commercially available architrave mouldings

The steps below and on page 191 show how to fit an architrave to a smooth surface. If the surface is not smooth, the back of the architrave may have to be scribed to fit to it before finally fixing in place. This is described in the section 'Scribing to walls'.

How to fit an architrave

Step 1 Architraves are kept back from the front edge of a frame by 6–10 mm, which is known as the margin. This is the line we work to when fitting architraves. Hence, as Step 1 shows, draw the margin on the front edge of the legs and head of the frame until they meet.

Step 2 With the heel (narrow point or inner edge) of the architrave to the margin, place the leg on the floor and mark the architrave where the margin marks intersect each other on the frame.

Figure 9.56 Step 1 Drawing a working margin

Figure 9.57 Step 2 Marking the architrave

Step 3 Place the architrave into a mitre box and cut using a tenon saw. Alternatively, use a combination square to mark a 45° angle on the architrave and cut freehand.

Step 4 Position the cut leg against the margin marks and fix with nails. (When possible try and nail through the **quirks** in the moulding to help hide the nails.) If the surface is uneven you may need to shape the back of the architrave to fit to it before finally fixing, as described in the section 'Scribing to walls'.

Remember

Use a block plane when planing end grain

Figure 9.58 Step 3 Mitring the architrave legs

Figure 9.59 Step 4 Fixing the architrave legs

Step 5 Cut a 45° angle on the head piece and offer it to the angle of the leg that has just been fixed. If you have a good fit, repeat Step 2 with the head and then the remaining leg. Should the mitres not go together, the head piece will need planing until it does.

Step 6 Punch all nails below the surface and remove the **arris** from the toe (wide part of the architrave – its outer edge).

Key term

Arris – the sharp edge formed when two flat or curved surfaces meet

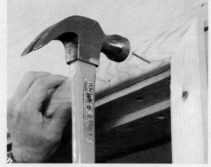

Figure 9.60 Step 5 Fixing the architrave head piece

Figure 9.61 Step 6 Finishing off the architrave

Scribing to walls

Scribing is to mark the profile of something onto the surface against which it is to be butted. In the case of architraves (or skirting) it is used so that the back can be shaped to fit against an uneven surface. The method shown below is one of a range of possible methods.

Step 1 Lightly nail the architrave 20 mm to 25 mm in front of the door frame or lining of the opening and parallel to it.

Step 2 Measure the distance between the front edge of the architrave and the margin mark.

Step 3 Cut a block, or set a pair of compasses, to the measurement you have just worked out. Take the architrave off the frame. Copy the shape of the uneven surface onto its face by running a compass, or block and pencil, down the uneven surface.

Step 4 Using a saw or a plane cut down to the line. Try in position, making sure the front of the architrave is level with the margin mark. Fix with nails.

Figure 9.62 Step 1 Temporarily fix forward of the door frame

Figure 9.63 Step 2 Measure to the margin mark

Figure 9.64 Step 3 Copy the surface shape onto the architrave

Figure 9.65 Step 4 Shape the architrave

Remember

Skirting boards may have to be fitted clear of the floor to fit block flooring, etc., so always check plans carefully

Safety tip

When fixing masonry nails, always use protective glasses

Key term

Dovetailed nails – pairs of nails angled in towards each other

Skirting board

Skirting boards are used to provide a decorative finish between floor and walls. They also protect the wall finish from damage. Skirting boards are available in lengths ranging from 1.8 m up to 6.3 m, increasing by multiples of 300 mm. They are between 75 mm and 175 mm deep and 19 mm to 25 mm thick.

There are only three joints to cut when fitting skirting boards:

- internal joints, which should be scribed
- external joints, which should be mitred
- angle-lengthening joints.

How to fit skirting

Skirting boards can be fixed by:

- adhesive
- masonry nails
- screws and plugs
- oval/lost head nails (for internal stud walls).

Before starting, remove all obstructions from the floor, such as lumps of plaster. Then check lengths of timber and choose the most economical lengths.

The boards should be fixed on the top edge and at the bottom and nails or screws must be made as inconspicuous as possible. Masonry nails should be **dovetailed**. Always plug the hole made by the screw or nail, using either a wooden plug or filler.

To get a board tight to the floor it is a good idea to use a kneeling board.

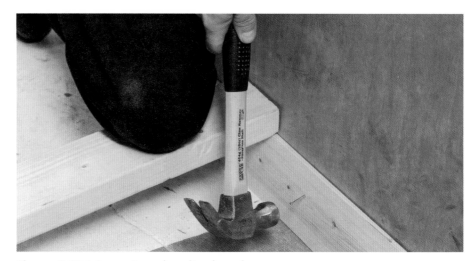

Figure 9.66 Dovetailed nails   Figure 9.67 Joiner using a kneeling board

Scribed joint

Skirting boards are shaped and frequently have mouldings to make them look attractive. Where they meet in a corner, the board butting up against the moulding needs to have its end shaped so that the boards fit neatly together. With one board fitted in place right into the corner, we now need to shape the board that will butt against it.

Method 1

Step 1 Place the square end of the board to be cut against the fixed board and copy the shape of the moulding onto the face of the board to be cut.

Step 2 Back cut with a coping saw and/or tenon saw to the line. Keep trying in place until a good joint has been achieved and fit in place.

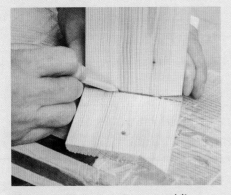

Figure 9.68 Step 1 Copy moulding shape to be scribed

Figure 9.69 Step 2 Back cut to the line

Method 2

Step 1 Using a mitre block, internally mitre across the width of the skirting board to be shaped. This will now show the end grain on the board.

Step 2 Using a coping saw, and/or tenon saw, back cut the mitre by letting the saw blade follow the line where the straight grain on the fixed board meets the end grain that is showing on the face of the board being shaped. Try in position and fit when a good joint has been achieved.

Figure 9.70 Step 1 Mitred skirting board with end grain to be shaped

Mitre joint

A completed mitre joint is shown in Figure 9.71. A step-by-step process for achieving this is as described below.

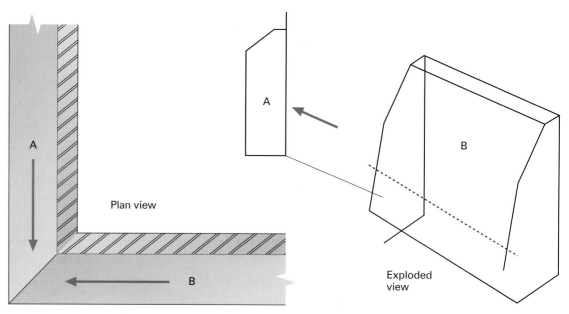

Plan view

Exploded view

Figure 9.71 Mitre joint on skirting board

Step 1 Lay the skirting board flat against the wall and draw a line on the floor where the board touches it until the line reaches the adjoining wall. Repeat this with the board flat against the other wall. When the board is removed the lines on the floor should meet at a point near to the angle of the wall.

Step 2 Place the stock of a sliding bevel right into the angle of the wall and move the blade around until it touches where the lines on the floor meet. You have just bisected the angle!

Step 3 Place the bevel on the top edge of the skirting board and draw the line for the mitre. Cut accurately. Try in position and fit when a good joint has been achieved. Any adjustments can be made using a block plane.

Figure 9.72 Step 1 Mark floor lines

Figure 9.73 Step 2 Bisect the angle

Figure 9.74 Step 3 Mark the mitre angle on the skirting board

Angle-lengthening joint

Angle-lengthening joints are used to join two pieces of board together in length. This will be required if the skirting board available is too short for the wall.

Simply cut a 45° angle across the width of one board and do the opposite cut on the board to which it will be joined. Put them together and **skew-nail** through the joint to hold it in position. The finished joint is shown in Figure 9.75.

Key term

Skew-nail – knocking nails in at angles so that the wood cannot be pulled away easily

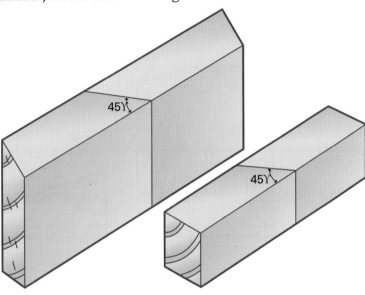

Figure 9.75 Angle-lengthening joint

Plinth block

Plinth blocks are also referred to as architrave blocks. They are fixed at the bottom of architraves and allow skirting to run up to them. See Figure 9.76.

They are used because:

- it is then possible to use a skirting board that is thicker than the architrave
- skirting can have a moulded back edge
- they provide protection to the moulding on the architrave
- they look elegant.

Plinth blocks should be approximately 15 mm taller than the skirting board and follow the basic profile of the architrave with at least a 6 mm excess.

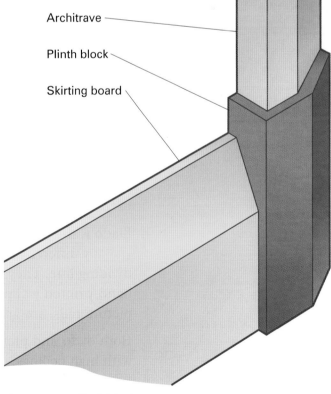

Architrave

Plinth block

Skirting board

Figure 9.76 Plinth block

Butt jointed – the simplest joint between two pieces of wood, with the end grain of one meeting the long grain of the other and glued, screwed or nailed together

They should be **butt jointed** to the architrave or preferably fixed using a barefaced tenon as shown in Figure 9.77. The skirting board can also be butt jointed to the plinth block, although a housing joint would be a better option, as shown in Figure 9.78. Mortise and tenon joints are covered in more detail in Unit 2034.

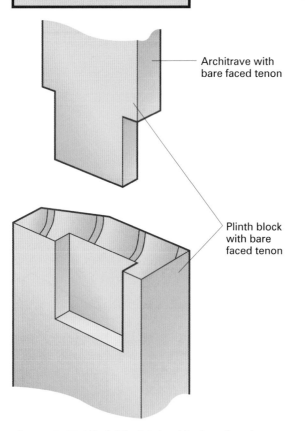

Architrave with bare faced tenon

Plinth block with bare faced tenon

Figure 9.77 Plinth block joined by barefaced tenon

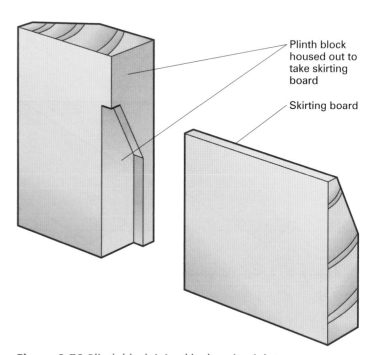

Plinth block housed out to take skirting board

Skirting board

Figure 9.78 Plinth block joined by housing joint

Picture rails and dado rails

Pictures can be hung from a picture rail using special clips that are fastened to the picture frame, but the rails are often now just used for effect.

Dado is the name used for the lower part of a wall when it is visually different from the upper part. Dado rails run along the dividing line. They were originally designed to protect expensive hand-printed wallpapers from chair and sofa backs but, like picture rails, are generally now used for effect.

Both dado and picture rails help reduce the visual height of rooms with high ceilings.

The same sequence and methods employed to fix skirting boards are used to fix dado and picture rails.

K3. Know how to install service encasements and cladding

This section has been designed to provide you with the knowledge and understanding to correctly encase a variety of services and structures.

The phrase 'encasing services' usually refers to the carcass, framework and trim that cover British standard beams (BSB), service pipes, cables, steel and concrete columns or, in some instances, unsightly spaces such as where a bath is situated or where the bulkhead for a staircase cuts into a room.

Whatever you are cladding there is a variety of material to choose from:

- plasterboard
- man-made timber boards such as plywood or MDF
- tongue and groove boarding
- melamine-faced boards.

The choice of materials depends on the finish wanted by the client and is requested in the specification.

In commercial areas, such as toilets in public buildings, cladding is usually done with melamine-faced boards. These boards protect the pipework from damage, but they are also chosen with hygiene in mind as they are easy to wipe clean. Access to the services behind is usually by way of removable panels which use brackets to hook into place.

> **Did you know?**
>
> BSB stands for British standard beam, often used to provide load-bearing supports in a building because the shape is resistant to bending

Figure 9.79 Urinals with removable panels

Figure 9.80 Urinals

Beams and columns

In some buildings BSBs have to be cladded to protect them from the effects of fire. The amount of protection required will depend on the function and location of the beam.

How to clad a beam

If a beam to be clad is made of timber, or even concrete, it can have a frame fixed to it and then facing material put on to the frame, or have the facing material fixed directly to it. The bigger problem is to clad a steel beam, often a BSB is used as a load-bearing support.

> **Key term**
>
> **Nogging** – a short length of timber, most often found fixed in a timber frame as a brace

Step 1 Fix **noggings** between the rolled edges of the beam. This can be done by accurately cutting lengths of 50 mm × 50 mm timber to match the vertical gap in the side of the BSB. Drive them vertically into the gap on both sides of the beam so they wedge in place. They will provide attachment points for a cradle so should be no more than 600 mm apart.

Alternatively, timber supports can be fitted in the gap along the whole length of the beam and then bolted into place through holes in the beam (some come with them or they can be drilled). These similarly provide attachment points for a cradle.

> **Key term**
>
> **Halving joint** – the same amount is removed from each piece of timber so that when fixed together the joint is the same thickness as the uncut timber

Step 2 Create a cradle using 50 mm × 25 mm treated softwood and use a simple **halving joint** (or housing or lap joint) to fix the corners together. Screw the cradle to the bearers. (If the noggings or timber supports come flush with the sides of the metal beam it is possible to dispense with a cradle by fixing additional timber supports along the beam near the top and bottom, as a support for facing material.)

Step 3 Run soffits along the length of the joist, fastening them into the cradle. The facing material can now be fixed to the soffits to conceal the sides of the beam.

Figure 9.81 Step 1 Fix noggings

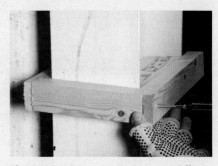

Figure 9.82 Step 2 Create a cradle

How to clad a column

Concrete or steel columns are cladded by constructing a set of framed grounds (either two or four) slightly larger than the column itself. These are generally shaped like a ladder and made from rough sawn timber. They are assembled around the column and adjusted until plumb.

Any slackness is taken up, using wedges or packing pieces with a screw or nail driven into them to stop them slipping.

Facing material is then fixed to the framed grounds.

Service pipes and cables

Service pipes and cables are hidden from view whenever possible. However, there are occasions when it is not possible to do so, for example when the circuit of pipe work has to go from one storey to another. When this occurs they should be encased behind a timber stud frame. The method is very similar to cladding a column.

Encasing pipes and wires

Figure 9.83 Step 3 Fitting soffits and facing material

Did you know?

Plumb means vertical; hence, a plumb bob is used to check whether something is vertical

Step 1 Measure how far the services protrude. The framework that is about to be fitted should not stand excessively from the object to be encased.

Step 2 By using halving joints or by butt nailing, construct two sets of framed grounds.

Step 3 Fix battens to the wall with plugs and screws or masonry nails.

Figure 9.84 Step 1 Measure the protrusion

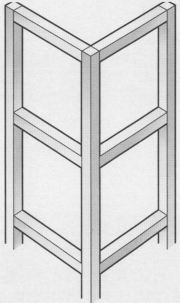

Figure 9.85 Step 2 Construct two sets of framed grounds

Figure 9.86 Step 3 Fix battens to the wall

Step 4 Fix the framed grounds to the battens.

Step 5 Clad with 6 mm ply or similar, scribing to the wall if necessary.

Figure 9.87 Step 4 Fix the framed grounds to the battens

Figure 9.88 Step 5 Clad with 6 mm ply

A corner casing for a large pipe is shown in Figures 9.89 and 9.90.

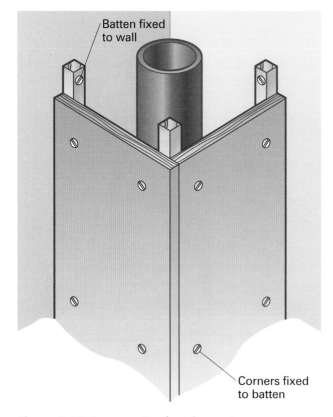

Batten fixed to wall

Corners fixed to batten

Figure 9.89 Corner casing for a large pipe

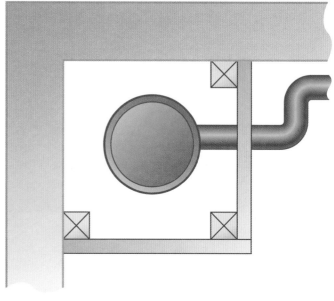

Figure 9.90 Plan view of casing

Alternative ways of encasing services are shown in Figures 9.91–9.94.

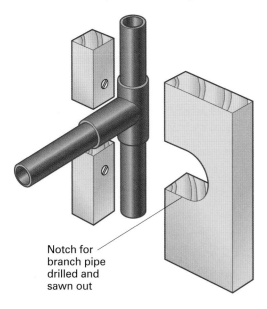

Notch for branch pipe drilled and sawn out

Figure 9.91 Branch pipe

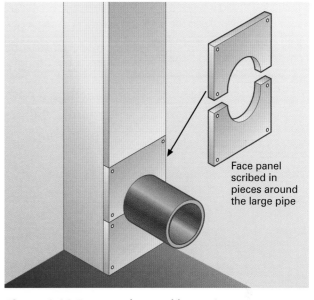

Face panel scribed in pieces around the large pipe

Figure 9.92 Face panel around large pipes

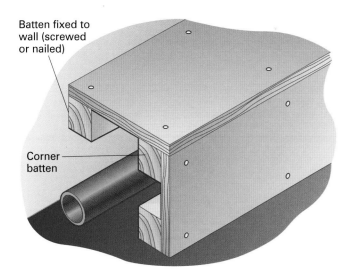

Batten fixed to wall (screwed or nailed)

Corner batten

Figure 9.93 Horizontal pipes

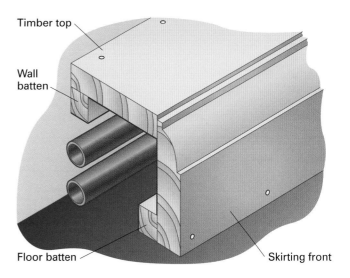

Timber top

Wall batten

Floor batten

Skirting front

Figure 9.94 Horizontal pipes with skirting board

One of the main dangers associated with drilling into walls or cutting into floors is the fact that there may be a hidden service such as electric cables, water pipes or even gas pipes behind the plaster or beneath the floorboards. There are tools designed to detect these services and they work by checking the area for the metal found in pipes and electric wires. More specialist tools can detect electric currents or the flow of water. Detection tools are not infallible and if you are in any doubt that the area you are cutting or drilling into has services, it is always safest to turn off the services at the mains.

Remember

Where there is a service valve, or stop cock, a removable panel must be fitted for maintenance and emergencies

Unit 2009

Know how to carry out second fixing

Baths

Baths are often panelled, but care should be taken to ensure that suitable materials are used that can cope with the heat and condensation associated with bathrooms. Also the plumbing should be easily accessible.

Panelling a bath is similar to encasing pipe work, only the framed grounds will probably be wider. Figures 9.95–9.97 illustrate typical framework and panel fixing.

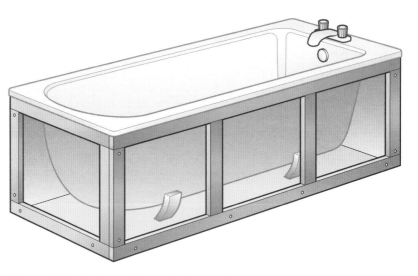

Figure 9.95 Framework for a bath panel

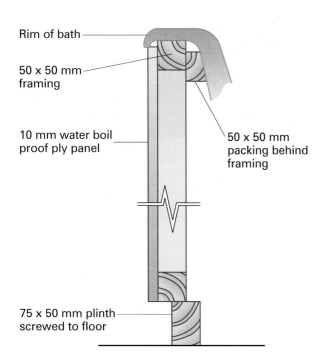

Rim of bath

50 x 50 mm framing

10 mm water boil proof ply panel

50 x 50 mm packing behind framing

75 x 50 mm plinth screwed to floor

Figure 9.96 Plinth below the frame as a toe space

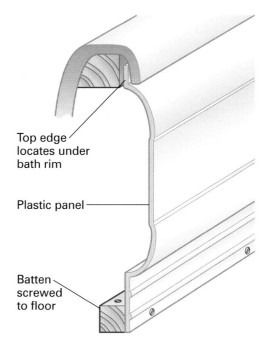

Top edge locates under bath rim

Plastic panel

Batten screwed to floor

Figure 9.97 Detail of a standard panel

Fixings that penetrate timber

Fixings are items of ironmongery that enable a carpenter to connect components. When choosing a fixing we must consider certain factors, which include:

- What strength must the fixing have?
- Where will the fixing be used?
- Will the fixing need to be removed at a later date?
- What is the cost?

The best sources for finding out what fixings are available are trade catalogues, local builders' merchants or DIY superstores. New types of fixing are regularly added to an already extensive range.

Although there are many specialist fixings available the most common are:

- nails
- screws
- wall fixings for solid walls
- wall fixings for hollow walls
- adhesives.

Nails

Nails consist of a head and shank and are inserted by a hammer or mechanical tool. There are several types, made from either ferrous or non-ferrous metal. Ferrous metal contains iron and will rust unless protected. The carpenter must decide on the most appropriate nail for the required application.

Round wire nails are available in sizes from 25 mm to 150 mm. They should not be driven below the surface of the timber and are relatively easy to remove. They are used for low-quality work where they will not be seen such as roofing and studwork.

Annular ring nails are available in sizes from 20 mm to 75 mm and are **sherardised** to prevent rusting. These nails are similar to the round wire nail, but feature a series of rings along the shank that makes them much harder to remove and also provides a stronger hold.

Did you know?

Some hardwoods are acidic and when unprotected ferrous metals are inserted the process of oxidisation (rusting) is accelerated and they stain the timber

Key term

Sherardised – a process of covering metal with zinc, a non-ferrous metal, to reduce rusting

Figure 9.98 Round wire nails

Figure 9.99 Annular ring nails

Figure 9.100 Oval wire nails

Figure 9.101 Panel pins

Figure 9.102 Lost head nails

Oval wire nails are available in sizes from 25 mm to 100 mm. They are manufactured from ferrous metal and can be punched below the surface of the timber. They are less likely to split the grain of the timber and are usually used for higher-quality work than the round wire nail.

Lost head nails are available in sizes from 40 mm to 75 mm. The head can be punched below the surface of the timber for concealment.

Panel pins are available in sizes from 20 mm to 40 mm. They are easy to punch below the surface, causing little damage to the face of the work. They are used for fine applications. Variations include sherardised and brass versions that resist rust and veneer pins for extra fine work.

Other nails include plasterboard, felt and clout nails, plastic-headed nails for use with uPVC systems and double-headed nails for shuttering work. All are designed for specific applications.

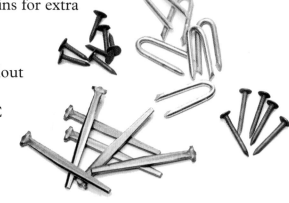

Figure 9.103 Other nails

Screws

Most modern screws are computer designed. Like a nail, screws consist of a head and a shank. However, the shank is threaded and designed to pull the fixing into the material into which it is being inserted.

Screws are manufactured from both ferrous and non-ferrous materials and are defined by:

- head type
- length, measured from the tip to the part of the head that will be flush with the work surface, ranging from 12 mm to 150 mm
- gauge, the diameter of the shank, ranging from 2 mm to 6.5 mm.

Once again, it is the carpenter's responsibility to choose the correct screw for the application in which it is being used.

Remember

Screws are still sold by the old Imperial Measures gauge number (8, 10, 12 etc.), but these are being phased out

Head types

Screws with countersunk heads are used when the screw has to be flush with the work or below it.

Raised head screws are usually used for attaching metal components such as door handles. Round heads are usually used for attaching sheet material to timber that is too thin to countersink.

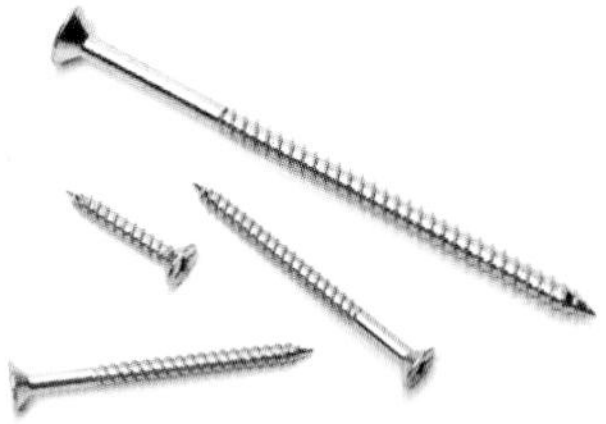

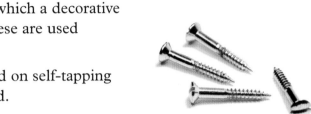

Figure 9.104 Countersunk

Mirror screws have a thread within the head to which a decorative dome can be attached. As the name suggests, these are used mainly for fixing mirrors.

Pan heads and flange heads are commonly found on self-tapping screws where the fixing of sheet metal is involved.

Figure 9.105 Raised head screws

Figure 9.106 Mirror screw head

Figure 9.107 Pan head

Figure 9.108 Flange head

Wall fixings for solid walls

Plug fixings

Nails and screws can be used to fix components to masonry. However, the carpenter must first plug the masonry, which can either be done using a plugging chisel or an electric drill.

If a plugging chisel is used carpenters can make their own plugs, but this is time-consuming and not commonly done. When using an electric hammer drill to plug a wall, several plug types are available:

- moulded plastic plugs
- hammer plugs
- frame fixings.

Figure 9.109 Moulded plastic plug

Figure 9.110 Hammer plug

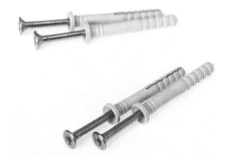

Figure 9.111 Frame fixings

All of these work on the same principle. A plastic segmented sleeve fits snugly into a hole that has been pre-drilled to the plug manufacturer's stated dimensions. A screw is then inserted into the plastic sleeve that pushes the segments apart to grip the side of the hole.

Anchor bolts

These are used for giving an extra strong fixing in concrete or masonry. They consist of a segmented metal sleeve that encases a conical plug on the end of a bolt. As the bolt is turned clockwise, the conical plug rises up the thread of the bolt, expanding the metal sleeve.

Figure 9.112 Anchor bolt

Wall fixings for hollow walls

Hollow wall fixings work on the principle of the fixing opening out behind the wall panel and gripping it in some way. These include:

- nylon anchors
- plastic collapsible anchors
- metal collapsible anchors
- rubber sleeve anchors
- gravity toggles
- spring toggles.

Figure 9.113 Nylon anchors

Figure 9.114 Plastic collapsible anchors

Figure 9.115 Metal collapsible anchors

Figure 9.116 Rubber sleeve anchors

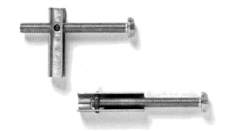

Figure 9.17 Gravity toggles

Figure 9.118 Spring toggles

Other fixings available are called EASI drivers™, but these are generally for use only in plasterboard, though they will support heavy items such as kitchen wall units or radiators. It is worth the effort to experiment with different types of fixing before deciding which one to use.

Adhesives

Adhesives, able to provide long-term fixing, are a relatively recent development within the building industry, but already there are numerous products available. These range from dry lining adhesive through to expanding foam and trade name products such as No Nails.

Datum points

The need to apply levels is required at the beginning of the construction process and continues right up to the completion of the building. The whole country is mapped in detail and the Ordnance Survey places datum points (bench marks) at suitable locations from which all other levels can be taken.

Ordnance bench mark (OBM)

Ordnance bench marks (OBM) are found cut into locations such as walls of churches or public buildings. The height of the OBM can be found on the relevant Ordnance Survey map or by contacting the local authority planning office. Figure 9.119 shows the normal symbol used, though it can appear as shown in Figure 9.120.

Safety tip

With adhesive, or any other chemical, carefully read all instructions before use

Functional skills

Using datum points will allow you to practice FM 1.2.1b relating to interpreting information from sources such as diagrams, tables, charts and graphs.

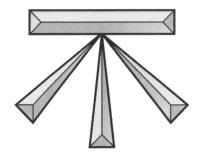

Figure 9.119 Ordnance bench mark

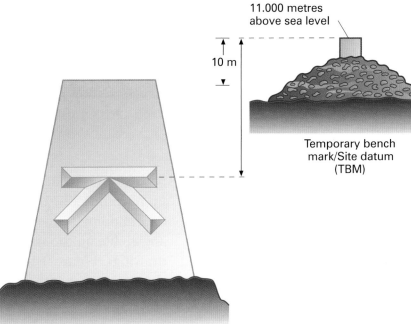

11.000 metres above sea level

10 m

Temporary bench mark/Site datum (TBM)

Ordnance Survey bench mark (OSBM) 10.000 metres above sea level

Figure 9.120 Site datum and OBM

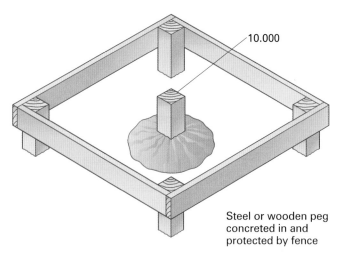

10.000

Steel or wooden peg
concreted in and
protected by fence

Figure 9.121 Datum peg suitably protected

Site datum

It is necessary to have a reference point on site to which all levels can be related. This is known as the site datum. The site datum is usually positioned at a convenient height, such as finished floor level (FFL).

The site datum itself must be set in relation to some known point, preferably an OBM, and must be positioned where it cannot be moved.

Figure 9.120 shows a site datum and OBM, illustrating the height relationship between them.

If no suitable position can be found a datum peg may be used, its accurate height is then transferred by surveyors from an OBM, as with the site datum. It is normally a piece of timber or steel rod positioned accurately to the required level and then set in concrete. However, it must be adequately protected and is generally surrounded by a small fence for protection, as shown in Figure 9.121.

Temporary bench mark (TBM)

When an OBM cannot be conveniently found near a site it is usual for a temporary bench mark (TBM) to be set up at a height suitable for the site. Its accurate height is transferred by surveyors from the nearest convenient OBM.

All other site datum points can now be set up from this TBM using datum points, which are shown on the site drawings. Figure 9.122 shows datum points on drawings.

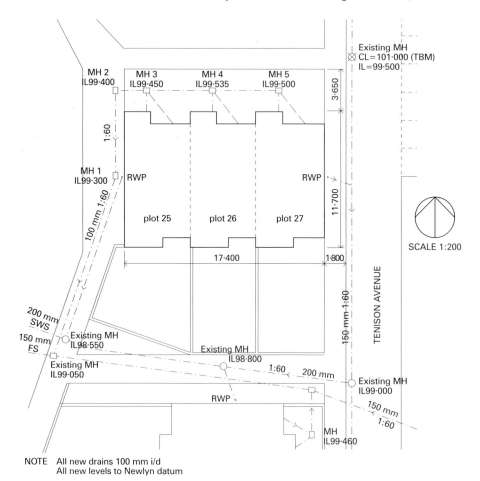

Figure 9.122 Datum points shown on a drawing

K4. Know how to install wall and floor units and fitments

This section is designed to provide you with the knowledge and understanding to assemble and fix both wall and floor units. Normally you will be fitting these in kitchens and bedrooms. Kitchens are the most difficult and are described below. If you learn to follow these instructions they will enable you to fit wall and floor units in a bedroom or anywhere else.

Kitchen safety

Most household accidents happen in the kitchen. Thus, when planning a kitchen, safety is the number one priority. Always:

- create a working triangle consisting of the sink, cooker and fridge, which avoids unnecessary movement and improves efficiency
- avoid a design that encourages people to use the working area of the kitchen as a cut through, to the garden or the living room for example
- try not to place a hob or sink at the end of a run of units and make sure that children cannot easily reach either the hob or the sink
- try to design work surfaces either side of the hob and avoid placing the hob in front of a window.

A plan of a well laid out kitchen is shown in Figure 9.123.

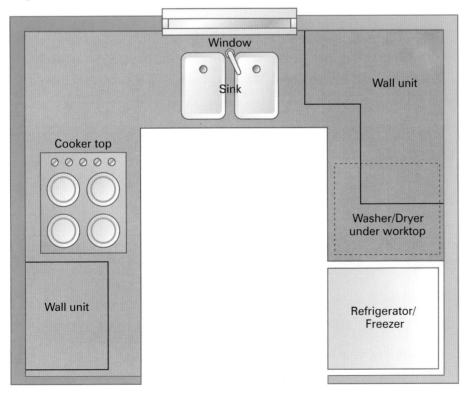

Figure 9.123 Kitchen plan

Kitchen units

The majority of kitchen units available today are constructed from melamine-faced particle board and supplied as either base units or wall units. Base units are floor standing, whereas wall units are mounted on the wall.

Units are supplied ready-built or as flat-pack. When supplied as flat-packs the assembly instructions must be followed with great care. Check whether shelves have to be fitted during assembly, rather than after installation, particularly in corner units.

Base units are available in widths of 300 mm to 1200 mm and are normally 600 mm deep and 900 mm high. Corner units are available to match them. Most manufacturers incorporate adjustable height legs in their designs to allow for uneven floors.

Wall units are available in widths of 300 mm to 1000 mm and are typically 600 mm, 720 mm or 900 mm high and 300 mm deep.

Worktops are also mainly made from plastic-laminated particle board and available in widths of 600 mm to 900 mm, but granite, stainless steel and proprietary, hard-wearing plastics are used in more expensive kitchens. The width fitted is dependent on the base unit used. Depth is usually 40 mm, but can vary. Plastic-laminated particle boards have a square- or post-formed front edge.

Fitting base units

> **Remember**
>
> When using a spirit level to mark a datum line, always alternate the level along its length; this counteracts inaccuracies and prevents cumulative error

Step 1 Carefully mark a horizontal datum line, approximately 1 m from the floor, on the walls that are to have base or wall units placed against them. We now have a point to work up or down from.

Figure 9.124 Step 1 Mark a horizontal datum line

Step 2 Determine a user-friendly height for the units minus the worktop. Then, working down from the 1 m datum, fix a back rail to the wall. We now have a datum that allows us to level across and along the units.

Step 3 Preferably working from a corner, place units where they should go, taking into consideration service pipes and cables. Adjust the legs so that all units are level and in line with the back rail.

Step 4 Once all units are in place they can be fixed together using connecting bolts and fixed to the wall or floor using screws.

Figure 9.125 Step 2 Fixing a back rail

Figure 9.126 Step 3 Adjust the unit's legs

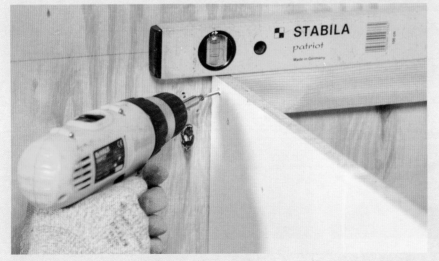

Figure 9.127 Step 4 Base units fixed in position

Unit 2009

Know how to carry out second fixing

Fitting worktops

Specialist tools and equipment are needed to cut and fix worktops. These include:

- jigsaw with downward cutting blades

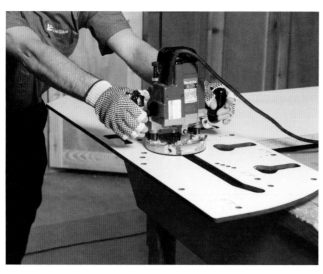

Figure 9.128 Joiner using router and jig

- plunge router (minimum 1600 watt) capable of plunging more than the depth of the worktop, with 12.7 mm diameter collet and template guide
- 12.7 mm diameter straight-cutting router bits, with the minimum length equal to the depth of the worktop
- worktop jig
- biscuit jointer
- clamps (G clamp or similar)
- contact adhesive
- varnish
- panel connectors.

Step 1 Cut to length and form an internal corner using a purpose-made jig. These are produced by a variety of manufacturers and, when used in conjunction with a powerful router, a clean accurate cut is achievable.

Step 2 Strengthen joints using a biscuit jointer.

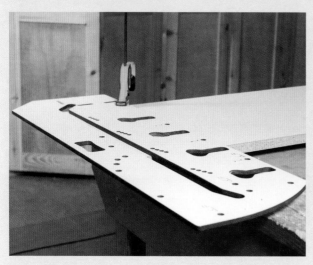

Figure 9.129 Step 1 Cut lengths and internal corners

Figure 9.130 Step 2 Strengthen joints with biscuit jointer

Step 3 Cut worktops to house sink units or appliances using a jigsaw with a downward cutting blade. This prevents the plastic-laminated surface from being chipped. The chipboard that is exposed by the cut should be coated in varnish to prevent any moisture penetration.

Step 4 Fix worktops to the base units with screws and connectors.

Figure 9.131 Step 3 Cut worktops to house sink units, etc.

Figure 9.132 Step 4 Fix worktops to base units

Fitting wall units

Most wall units are fastened to the wall by an adjustable bracket that hooks itself onto a steel hanger plate. The plans should show the clearance required between the worktop and base of the wall units. If not, check with a supervisor or the customer. The tops of all units are usually the same height.

Step 1 Measure the height of units and add this to the clearance from the work surface, which gives the height where the tops of the wall units should go. Measure and mark these, working upwards from the datum line.

Step 2 Mark the locations for the hanger plates and fix in place. Often the manufacturer of the wall units provides a template to help position them.

Step 3 Hang the wall units on the hanger plates and adjust for height by turning a screw housed within the adjustable bracket. Once done, another screw enables the bracket to be tightened on to the steel plate.

Figure 9.133 Step 1 Mark the tops of wall units

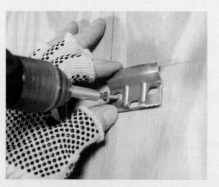

Figure 9.134 Step 2 Mark and fix hanger plates

Figure 9.135 Step 3 Hang and adjust wall units

As with base units, wall units can be connected by using connecting bolts.

Figure 9.136 Finish the job by fixing plinths, cornices, doors and shelves

Finishing off

Complete the job by fixing plinth boards, cornices, pelmets, doors, shelves and drawers. Details of how to do this should be provided with the kitchen units and often vary between manufacturers.

In general:

- plinth boards are fixed to the legs of base units using clips supplied by the manufacturer. However, the boards may need scribing to the floor to avoid having any unsightly gaps

- cornices are often fitted on to wall units. This requires accurate marking out and either a chop saw or mitre saw must be used to give a clean, accurate cut. Care must be taken to ensure the cornice does not move or slip while it is being cut

- doors and shelves can be fitted to the units. This should be done last to avoid damage to the face of the work. The runners that hold the shelves, and the hinges that hold the doors, both allow for adjustment. This ensures that an equal gap is seen between doors and shelves when the work has been completed.

Figure 9.137 A completed kitchen

Functional skills

When dealing with installation problems onsite you will have the opportunity to practice the *Interpreting* elements of functional skills, e.g. FM 1.3.1 Judge whether findings answer the original problem, FM 1.3.2 Communicate solutions to answer practical problems.

Working life

Shauna has been asked to fit 300 mm x 30 mm skirting boards in a room. The architrave is 75 mm x 19 mm.

What might be a problem at the junction where the architrave and the skirting board meet? How could Shauna get around this? The room is also to have a dado rail fitted. The rail is 5 mm thicker than the architrave. How could Shauna resolve this problem?

FAQ

What is a scribe?

A scribe is a copy of the surface it fits over.

What does bisecting an angle mean?

Cutting an angle equally in two, as when cutting a mitre to create an internal angle.

Do I scribe both ends of a length of skirting?

No, it looks neater if one end butts to the wall and the other end is scribed to the butted end of the next skirting.

Can I have a scribe and a mitre?

No, it would be very difficult to get the scribed edges to meet correctly.

Check it out

1   Name three examples of mouldings.
2   Explain how an architrave can be fitted to a surface that is not smooth.
3   Describe the different ways in which skirting can be fitted.
4   Briefly describe the two different methods of shaping skirting where it meets in a corner.
5   Explain why plinth blocks are used.
6   Describe in your own words the following types of door:
    a framed door
    b flush door
    c fire resisting door.
7   Why should a door be stored flat for a few days prior to fitting?
8   If a door is too tall for an opening, where should you cut excess wood from?
9   State the size, type and number of hinges usually fitted to an external door.
10 Explain the purpose of a spy hole.
11 Explain what push plates are used for.
12 Explain why a door closer should be fitted to a door.
13 Explain the special feature of a helical or double action hinge.
14 Why might a rolled steel joist (RSJ) need to be cladded?
15 What should you always consider when encasing service pipes or cables?
16 When fitting kitchen wall units, how do you know how much clearance is required between the worktop and the base of the unit?
17 Why are the doors and shelves of kitchen units fitted last?

Getting ready for assessment

The information contained in this unit, as well as continued practical assignments that you will carry out in your college or training centre, will help you with preparing for both your end of unit test and the diploma multiple-choice test. It will also aid you in preparing for the work that is required for the synoptic practical assignments.

The information contained within this unit will aid you in learning how to identify and calculate the materials and equipment required for second fixing.

You will need to be familiar with:

- preparing the materials for a range of second fixing tasks
- installing side hung doors and ironmongery
- installing mouldings
- installing service encasements and cladding
- installing wall, floor units and fitments.

This unit will have made you familiar with the common second fixing tasks. For example, for learning outcome four you have seen the methods used for fixing wall and floor units, whilst protecting electric, gas and water pipes. You have also seen the importance of using drawings and specifications to complete work accurately. You will need to follow this information in order to successfully assemble and install wall and floor units. You will also need to select the correct tools to complete the work to specification.

The knowledge from this unit will also need to be used to install worktops, form openings for hobs and sinks and returned post-formed worktops using worktop jigs. This work will always need to be of a high standard, as it is client-facing and will leave the client with an overall impression of the quality of your work.

Before you start work on the synoptic practical test it is important that you have had sufficient practice and that you feel you are capable of passing. It is best to have a plan of action and a work method that will help you. You will also need a copy of the required standards, any associated drawings and sufficient tools and materials. It is also wise to check your work at regular intervals. This will help you to be sure that you are working correctly and help you to avoid problems developing as you work.

Your speed at carrying out these tasks will also help you to prepare for the time limit that the synoptic practical task has. But remember, don't try to rush the job as speed will come with practice and it is important that you get the quality of workmanship right.

Always make sure that you are working safely throughout the test. Make sure you are working to all the safety requirements given throughout the test and wear all appropriate personal protective equipment. When using tools, make sure you are using them correctly and safely.

Good luck!

CHECK YOUR KNOWLEDGE

Unit 2009
Know how to carry out second fixing

1 Architraves should be kept back from the front edge of a frame by:
  a 4–6 mm.
  b 6–10 mm.
  c 10–12 mm.
  d 12–14 mm.

2 The moulding that is fixed at the bottom of an architrave where it meets the skirting is called a:
  a plinth block.
  b pelmet block.
  c cornice block.
  d dado block.

3 What is a door stile?
  a the longest vertical timber in the door frame.
  b the longest horizontal timber in the door frame.
  c the shortest vertical timber in the door frame.
  d the shortest horizontal timber in the door frame.

4 Which of these is not a standard door height?
  a 1.97 m.
  b 2 m.
  c 2.03 m.
  d 2.17 m.

5 What could cause a door to spring open when it is closed (binding)?
  a the stops are not fitted correctly.
  b the door and lining are fitted out of level.
  c the screws in the hinges are not in straight.
  d any of the above.

6 A lock block is contained in a:
  a flush door.
  b fire door.
  c panel door.
  d framed door.

7 Which hinge would you use if a door is to be fitted in an area with a floor that slopes upwards?
  a tee hinge.
  b rising butt hinge.
  c storm proof hinge.
  d butt hinge.

8 What ironmongery should be fitted to a fire escape door?
  a spy hole.
  b letter plate.
  c panic bar.
  d dead lock.

9 Ferrous nails should not be used outside because:
  a they will rust.
  b they are weaker.
  c they are not long enough.
  d they are too long.

10 Which type of screw head would be used for fixing metal components such as door handles?
  a mirror screw.
  b flange head.
  c countersunk.
  d raised head.

11 When it is not possible to hide service pipes and cables from view (for example when the circuit of pipe work has to go from one storey to another), what should you encase the services behind?
  a a tiled screen.
  b a timber stud frame.
  c an easily shattered covering.
  d nothing – they must stay visible.

12 The best way to cut a worktop joint is to use a:
  a jig saw.
  b router and template.
  c hand saw.
  d circular saw.

How to erect structural carcassing

While structural carcassing may sound complex, it is the skeleton of the structure you are building. The term covers any form associated with a building's structure. This unit covers the key areas of structural carcassing you are likely to encounter when working at this level. By the end of the unit, you will be confident in performing the tasks described in the following pages.

This unit also supports NVQ Unit VR11 Erect Structural Carcassing Components.

This unit contains material that supports TAP unit 4 Erect Structural Carcassing Components. It also contains material that supports the delivery of the five generic units.

This unit will cover the following learning outcomes:

- How to erect truss rafter roofs
- How to construct gables, verge and eaves
- How to install floor joists.

Basic roofing

Roofing is a complex aspect of the construction industry and it is wise to begin with a basic introduction to roofing, particularly terminology and roof types, many of which you will have seen before, but few of which you will be able to name! By the end of this section, you will have the necessary background to tackle basic roof types with confidence.

Roofing terminology

Roofs are made up of a number of different parts called 'elements'. These in turn are made up of 'members' or 'components'.

Elements

The main elements, shown in Figure 10.1, are:

- **gable** – the triangular part of the end wall of a building that has a pitched roof
- **hip** – where two external sloping surfaces meet
- **valley** – where two internal sloping surfaces meet
- **verge** – where the roof overhangs at the gable
- **eaves** – the lowest part of the roof surface where it meets the outside walls.

Members or components

The main members or components, shown in Figure 10.1, are:

- **ridge board** – a horizontal board at the apex acting as a spine, against which most of the rafters are fixed
- **wall plate** – a length of timber placed on top of the brickwork to spread the load of the roof through the outside walls and give a fixing point for the bottom of the rafters
- **rafter** – a piece of timber that forms the roof, of which there are several types
- **common rafters** – the main load-bearing timbers of the roof
- **hip rafters** – used where two sloping surfaces meet at an external angle, this provides a fixing for the jack rafters and transfers their load to the wall
- **crown rafter** – the centre rafter in a hip end that transfers the load to the wall
- **jack rafters** – these span from the wall plate to the hip rafter, enclosing the gaps between common and hip rafters and crown and hip rafters

- **valley rafters** – like hip rafters but forming an internal angle, acting as a spine for fixing cripple rafters
- **cripple rafters** – similar to a jack rafter, these enclose the gap between the common and valley rafters
- **purlins** – horizontal beams that support the rafters mid-way between the ridge and wall plate.

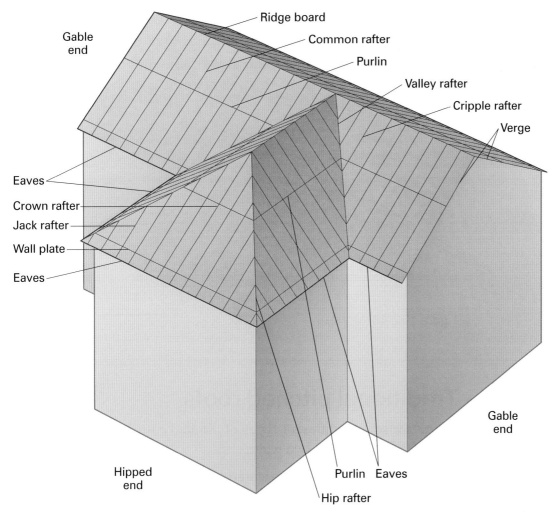

Ridge board
Gable end
Common rafter
Purlin
Valley rafter
Cripple rafter
Verge
Eaves
Crown rafter
Jack rafter
Wall plate
Eaves
Hipped end
Purlin Eaves
Hip rafter
Gable end

Figure 10.1 Roofing terminology

Basic terms for setting out

Setting out covers all the action necessary before commencing construction. Here are the most common terms you will need:

- **span** – the distance measured in the direction of the ceiling joists, from the outside of the wall to the wall plate, known as the overall span
- **run** – equal to half the span
- **apex** – the peak or highest part of the roof

Did you know?

The span, run and rise are measured in metric units, but the pitch is measured in degrees

- **rise** – the distance from the outside of the wall plates at wall-plate level to the apex
- **pitch** – the angle or slope of the roof, calculated from the rise and the run
- **pitch line** – a line that is marked up from the underside of the rafter, one third of its depth to the top of the birdsmouth cut
- **plumb cut** – the angle of cut at the top of the rafter
- **seat cut** – the angle of cut at the bottom of the rafter
- **birdsmouth** – notch cut out at the bottom of the rafter to allow the rafter to sit on the wall plate.

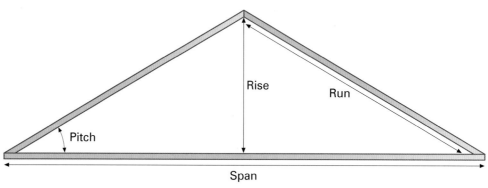

Figure 10.2 Rise and span

Traditional pitched roofs

There are several different types of pitched roof, but most are constructed in one of two ways:

- **trussed roof** – a prefabricated pitched roof specially manufactured prior to delivery on site, saving timber as well as making the process easier and quicker. Trussed roofs can also span greater distances without the need for support from intermediate walls
- **traditional roof** – a roof entirely constructed on site from loose timber sections using simple jointing methods.

Roof types

A pitched roof can be constructed either as a single roof, where the rafters do not require any intermediate support, or a double roof where the rafters are supported. Single roofs are used over a short **span**, such as a garage; double roofs are used to span a longer distance, such as a house or factory.

Key term

Span – the distance measured in the direction of ceiling joists, from the outside of one wall plate to another; known as the overall (O/A) span

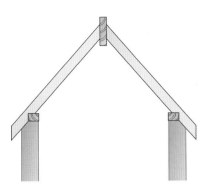

Figure 10.3 Single roof

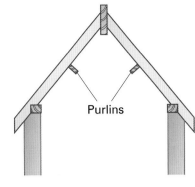

Figure 10.4 Double roof

There are many different types of pitched roof, including:

- **mono pitch** – with a single pitch
- **lean-to** – with a single pitch, which butts up to an existing building

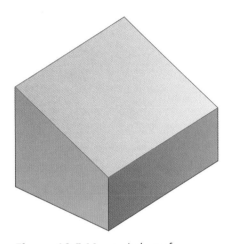

Figure 10.5 Mono pitch roof

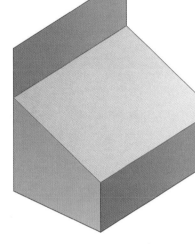

Figure 10.6 Lean-to roof

- **duo pitch** – with gable ends
- **hipped roof** – with hip ends incorporating crown, hip and jack rafters

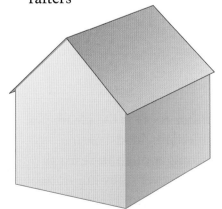

Figure 10.7 Duo pitch roof with gable ends

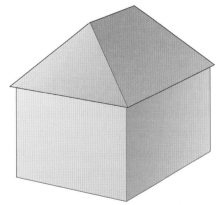

Figure 10.8 Hipped roof

- **over hip** – with gable ends, hips and valleys incorporating valley and cripple rafters
- **mansard** – with gable ends and two different pitches used mainly when the roof space is to be used as a room
- **gable hip** or **gambrel** – double-pitched roof with a small gable (gablet) at the ridge and the lower part a half-hip
- **jerkin-head** or **barn hip** – double-pitched roof hipped from the ridge part-way to the eaves, with the remainder gabled.

The type of roof used will be selected by the client and architect.

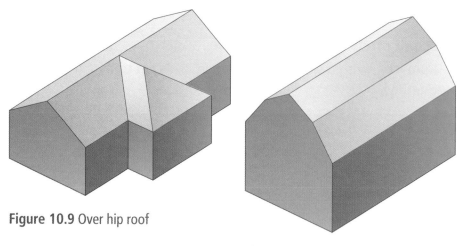

Figure 10.9 Over hip roof

Figure 10.10 Mansard roof

K1. Know how to erect truss rafter roofs

> **Key terms**
>
> **p.a.r.** – a term used for timber that has been 'planed all round'
>
> **Gang-nailed** – galvanised plate with spikes used to secure butt joints

Most roofing on domestic dwellings comprises factory made trussed rafters. These are made of stress graded, **p.a.r.** timber to a wide variety of designs, depending on requirements. All joints are butt jointed and held together with fixing plates, the face fixed on either side. These plates are usually made of galvanised steel and are either nailed or factory pressed. They may also be **gang-nailed** gusset plates made of 12 mm resin bonded plywood.

Trussed rafters

One of the main advantages of truss rafter roofs is the clear span achieved, as there is no need for intermediate, load-bearing partition walls. Standard trusses are strong enough to resist the eventual load of the roofing materials. However, they are not able to withstand pressures applied by lateral bending. Hence, damage is most likely to occur during delivery, movement across site, site storage or lifting into position.

Wall plates are bedded as described above. Following this, the positions of the trusses can be marked at a maximum of 600 mm between centres along each wall plate. The sequence of operations then varies between gable and hipped roofs.

Step 1 Fix first truss using framing anchors 1.

Step 2 Stabilise and plumb first truss with temporary braces E.

Step 3 Fix temporary battens on each side of ridge A.

Step 4 Position next truss 2.

Step 5 Fix the wall plate and temporary battens A. Continue until last truss is positioned.

Step 6 Fix braces B.

Step 7 Fix braces C.

Step 8 Fix horizontal restraint straps at max 2 m centres across trusses on to the inner leaf of the gable walls.

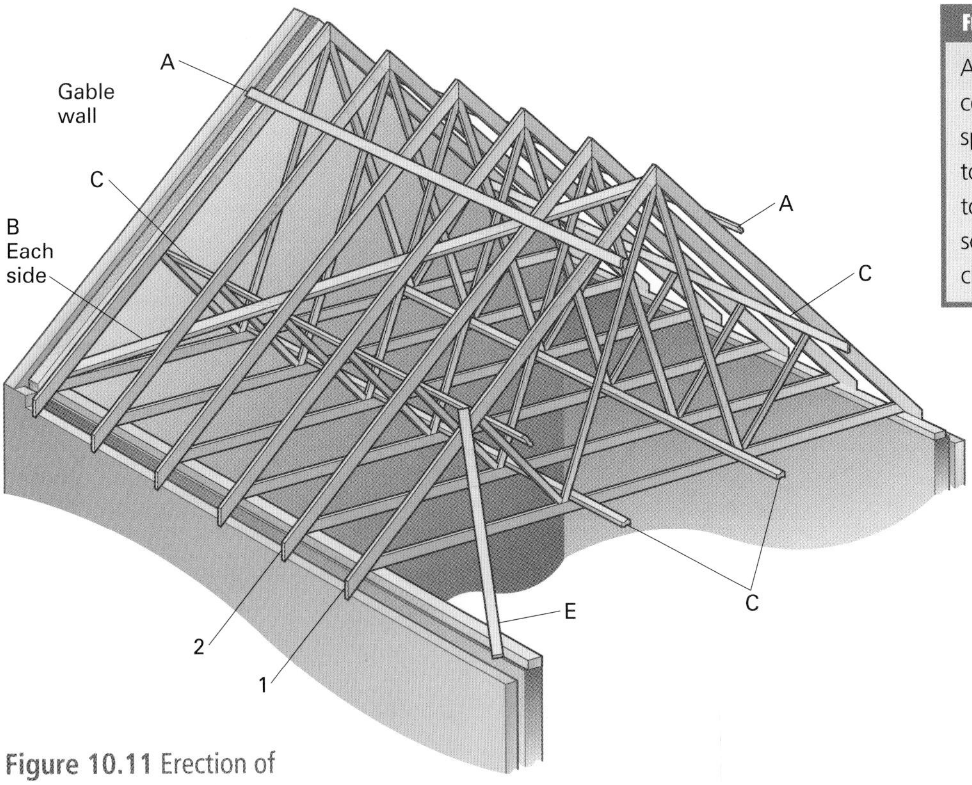

Figure 10.11 Erection of common trussed rafters

Trimming roof openings

Roofs often have components such as chimneys or roof windows. These components create extra work as the roof must have an opening for them to be fitted. **Trimming an opening** involves cutting out parts of the rafter and putting in extra support to carry the weight of the roof over the missing rafters.

Unit 2010 — How to erect structural carcassing

Chimneys

Chimneys are rarely used in new house construction as there are more efficient and environmental ways of heating these days, but older houses will have chimneys and these roofs must be altered to suit.

When constructing such a roof, the chimney should already be in place, so you should cut and fit the rest of the roof, leaving out the rafters where the chimney is. When you mark out the wall plate, make sure that the rafters are positioned with a 50 mm gap between the chimney and the rafter. You may also need to put in extra rafters.

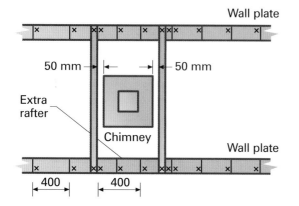

Figure 10.12 Wall plate marked out to allow for chimney

Next, fit the trimmer pieces between the rafters to bridge the gap, then fix the trimmed rafters – rafters running from wall plate to trimmer and from trimmer to ridge. The trimmed rafters are birdsmouthed at the bottom to sit over the wall plate, and the plumb and seat cut is the same as for common rafters.

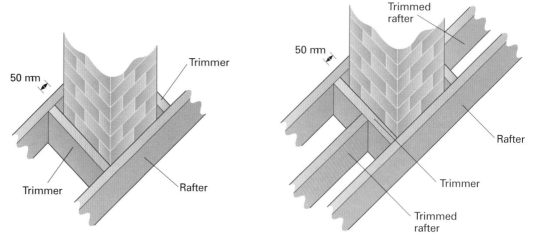

Figure 10.13 Opening trimmed around chimney with trimmers fitted

Key terms

Mid-pitch – in the middle of the pitch rather than at the apex or eaves

If the chimney is at **mid-pitch** rather than at the ridge, you will need to fit a chimney gutter. This ensures the roof remains watertight by preventing the water gathering at one point. The chimney gutter should be fixed at the back of the stack.

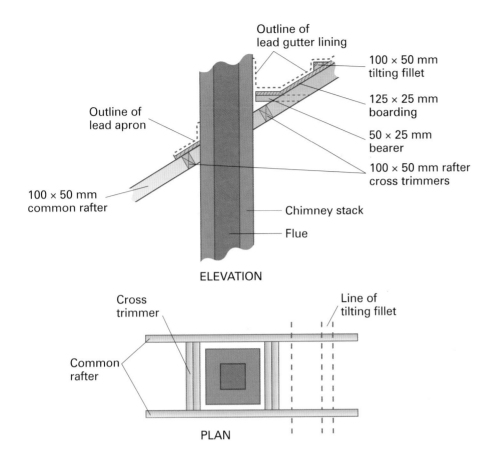

Figure 10.14 Gutter detail around chimney

Roof windows and skylights

You also need to trim openings for roof windows and skylights.

If the roof is new, you can plan it in the same way as for a chimney, remembering to make sure the area you leave for trimming is the same size as the window.

If you need to fit a roof window in an existing roof, the procedure is different.

Step 1 Strip all the tiles or slates from the area where you want to fit the window.

Step 2 Remove the tile battens, felt and roof cladding.

Step 3 Mark out the position of the window and cut the rafters to suit.

Step 4 Fit the trimmer and trimmed rafters – you may need to double up the rafters for additional strength – and fit the window.

Step 5 Re-fit the cladding and felt, and fix any flashings.

Step 6 Cut and fit new tile battens and finally re-tile or re-slate the roof.

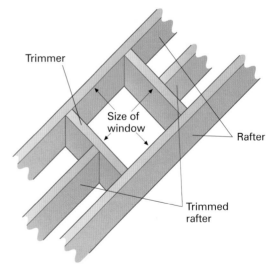

Figure 10.15 Roof trimmed for roof window or skylight

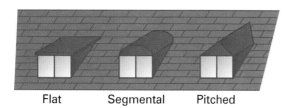

Flat Segmental Pitched

Figure 10.16 Dormer types

Figure 10.17 Framing for dormer

Dormers

Dormer windows are different to roof windows: a roof window lays flat against the roof, while a dormer window projects up from the sloping surface. A dormer window is often preferred to a roof window when there is limited headroom.

The construction of a dormer is similar to that of a roof window, except once the opening is trimmed, you need to add a framework to give the dormer shape. For a dormer you usually double up the rafters on either side to give support for the extra load the dormer puts on the roof.

Transporting and lifting truss rafters

The transportation, lifting and storing of trussed rafters is important in order to prevent damage.

Trusses should be stored either upright in a stack or laid flat and supported on bearers. When trusses need to be lifted it is vital that they are well supported and that enough people are available to lift them safely.

Constructing truss rafter roofs at ground level

An alternative way of constructing a truss roof is to construct the roof on the ground first. This can be done on site if there is enough space, but usually they are made in a factory and delivered by lorry. This method of construction removes the risk of height while fixing the trusses, but does not eliminate it as the completed roof has to be fixed once it is craned into place.

Figure 10.18 Truss roof lifted by a crane

Valleys

The next section deals with valleys, which are formed when two sloping parts of a pitched roof meet at an internal corner.

Marking out for valleys

Valleys can be worked out in the same way as hips, using either a ready reckoner or geometry. Here we will look at geometry.

Working out the angles for valleys is similar to doing so for hips, except that the key drawing is not a triangle, but a plan drawing of the roof.

First, you need to find out the valley rafter true length, plumb and seat cut. Start by finding the rise of the roof and drawing a line this length at a right angle to the valley where it meets the ridge. Join this line to the point where the valley meets the wall plate. This will give you the true length of the valley rafter as well as the plumb and seat cuts.

As with the hip rafter, there are two other angles to find for a valley rafter: the dihedral angle and the edge cut.

The dihedral angle for the valley is used in the same way as the hip dihedral and again rarely in roofing today. Figure 10.21 shows you how to work out the dihedral angle.

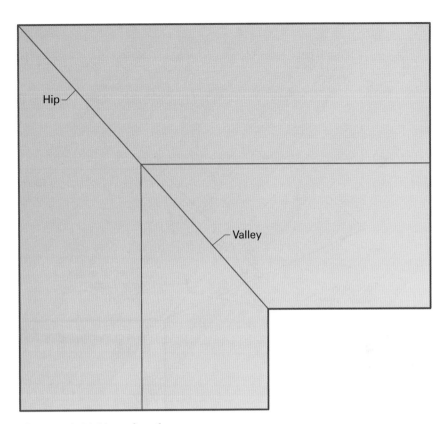

Figure 10.19 Plan of roof

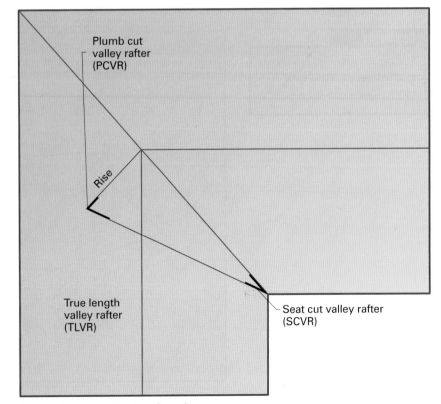

Figure 10.20 Valley true length

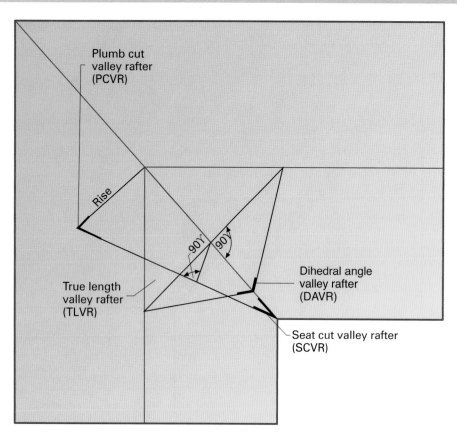

Figure 10.21 Dihedral angle hip

The final angle to find is the edge cut for the valley rafter, as follows:

Step 1 Mark on the rise and true length of the valley rafter.

Step 2 Draw a line at right angles to the valley where it meets the wall plate and extend this line to touch the ridge at A.

Step 3 Set your compass to the true length of the valley and **swing an arc** towards the ridge at B.

Step 4 Join up line A–B to give you the edge cut.

Key terms

Swing an arc – set a compass to the required radius and lightly draw a circle or arc

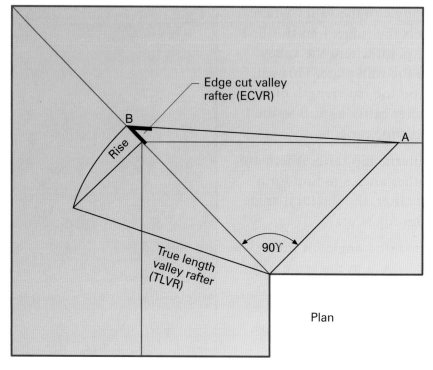

Figure 10.22 Edge cut

The final part of valley geometry is to find the true length and edge cut for the cripple rafters, as follows:

Step 1 Draw out the roof plan as usual, then to the side of your plan draw out a section of the roof.

Step 2 Set your compass to the rafter length and swing an arc downwards.

Step 3 Draw line A downwards until it meets the arc, then draw a line at right angles to line A until it hits the wall plate, creating line B.

Step 4 Draw a line from where line B hits the wall plate up to where the valley meets the ridge. This will give you the appropriate true length (TLCrR) and edge cut (ECCrR).

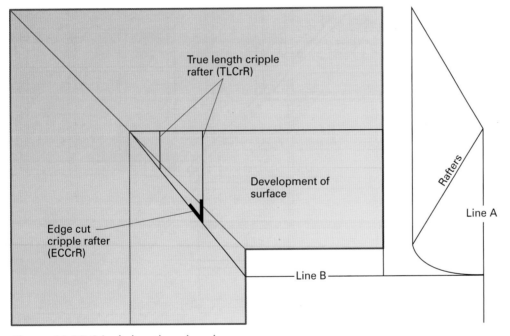

True length cripple rafter (TLCrR)

Development of surface

Edge cut cripple rafter (ECCrR)

Line B

Rafters

Line A

Figure 10.23 Cripple length and angle

Setting out a valley

There are four steps to follow when setting out a valley.

Step 1 Fit the wall plate and mark it out with the position of the common rafters.

Step 2 Fit the common rafters and ridge.

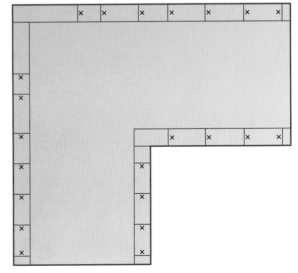

Figure 10.24 Step 2 Common rafters fitted

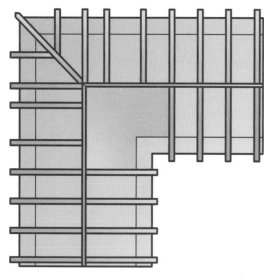

Figure 10.25 Step 3 Hip and jack rafters fitted

Key term

Lay board – a piece of timber fitted to the common rafters of an existing roof to allow the cripple rafters to be fixed

Step 3 Fit the hip and jack rafters.

Step 4 Fit the valley and cripple rafters, taking the true lengths and bevels from the drawings.

An alternative to using valley rafters is to use a lay board. **Lay boards** are most commonly used with extensions to existing roofs, or where there are dormer windows.

The lay board is fitted onto the existing rafters at the correct pitch, then the cripple rafters are cut and fixed to it.

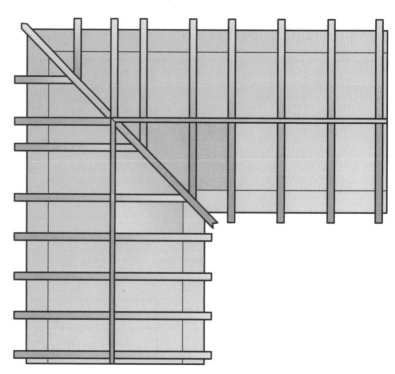

Figure 10.26 Step 4 Valley and cripple rafters fitted

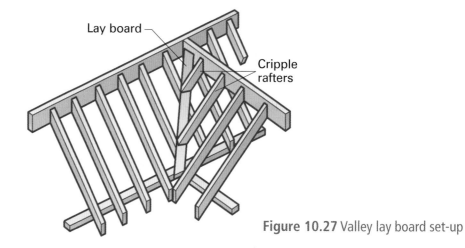

Lay board

Cripple rafters

Figure 10.27 Valley lay board set-up

Roof coverings

Once all the rafters are on the roof, the final thing is to cover it. There are two main methods of covering a roof, each using different components. Factors affecting the choice of roof covering include what the local weather is like and what load the roof will have to take.

Method 1

This method is usually used in the north of the country where the roof may be expected to take additional weight from snow.

Step 1 Clad the roof surface with a man-made board such as oriented strand board (OSB) or exterior grade plywood.

Step 2 Cover the roof with roofing felt starting at the bottom and ensuring the felt is overlapped to prevent water from getting in.

Step 3 Fit the felt battens (battens fixed vertically and placed to keep the felt down while allowing ventilation) and the tile battens (battens fixed horizontally and accurately spaced to allow the tiles to be fitted with the correct overlap).

Step 4 Finally, fit the tiles and cement on the ridge.

Method 2

This is the most common way of covering a roof.

Step 1 Fit the felt directly onto the rafters.

Step 2 Fit the tile battens at the correct spacing.

Step 3 Fit the tiles and cement on the ridge.

Another way to cover a roof involves using slate instead of tiles. Slate-covered roofing is a specialised job as the slates often have to be cut to fit, so roofers usually carry this out.

Ready reckoner

A **roofing ready reckoner** is a book used as an alternative to the geometry method and is often the simplest way of working out lengths and bevels. The book consists of a series of tables that are easy to follow once you understand the basics.

To use the ready reckoner you must know the span and the pitch of the roof.

Take a hipped roof with a 36 degree pitch and a span of 8.46 m. First you halve the span, getting a run of 4.23 m.

Referring to the tables in the ready reckoner, you can work out the lengths of the common rafter as follows:

Rise of common rafter 0.727 m per metre run			Pitch 36 degrees
Bevels:		Common rafter	Seat 36
		Common rafter	Plumb 54
		Hip or valley	Seat 27
		Hip or valley	Ridge 63
		Jack rafter	Edge 39
Jack rafters:		333 mm centres decrease	412 mm
		400 mm centres decrease	494 mm
		500 mm centres decrease	618 mm
		600 mm centres decrease	742 mm

Run of rafter	0.1	0.2	0.3	0.4	0.5	0.6	0.7	0.8	0.9	1.0
Length of rafter	0.124	0.247	0.371	0.494	0.618	0.742	0.865	0.989	1.112	1.236
Length of hip	0.159	0.318	0.477	0.636	0.795	0.954	1.113	1.272	1.431	1.590

You can see that the length of a rafter for a run of 0.1 m is 0.124 m, therefore for a run of 1 m, the rafter length will be 1.24 m – but you need to find the length of a rafter for a run of 4.23 m. This is how:

$$1.00 \text{ m} = 1.24 \text{ m}$$
$$4.00 \text{ m} = 4.994 \text{ m}$$
$$0.20 \text{ m} = 0.247 \text{ m}$$
$$0.03 \text{ m} = 0.037 \text{ m}$$
$$\text{Total} = 5.228 \text{ m}$$

The length of the common rafter is 5.228 m. However, there are a few adjustments you must make before finding the finished size.

You need to allow for an overhang and for the ridge, which can both easily be measured. For the purposes of this example, we will use an overhang of 556 mm and a ridge of 50 mm. The final calculation is:

$$\text{Basic rafter} = 5.228 \text{ m}$$
$$+ \text{ overhang} = 0.556 \text{ m}$$
$$- \text{ half ridge} = 0.025 \text{ m}$$
$$\text{Total} = 5.759 \text{ m}$$

The rafter length is 5.759 m.

Now you need to refer back to the table, which tells you that, for the common rafter, the seat cut is 36 degrees and the plumb cut is 54 degrees.

You can now mark out and cut your pattern rafter as before, remembering to mark on the pitch line and plumb cut. Measure the original size (5.228 m for our example) along the plumb cut, mark out the seat cut and cut.

The hip and valley rafters are worked out in the same way, but jack and cripple rafters are different.

Jack and cripple rafters are marked on the table and are easy to work out. Continuing with the example, where the rafter length is 5.228 m, you look at the table and, if you are working at spacing the rafters at 400 mm centres, you reduce the length of the common rafter by 494 mm. The first jack/cripple rafter will be 5.228 m – 0.494 m = 4.734 m; the next jack/cripple rafter will be 4.734 m – 0.494 m = 4.240 m and so on. The angles for the rest of the cuts are all shown on the table.

Bevels:	Common rafter	Seat 36
	Common rafter	Plumb 54
	Hip or valley	Seat 27
	Hip or valley	Ridge 63
	Jack rafter	Edge 39

Jack rafters:	333 mm centres decrease	412 mm
	400 mm centres decrease	494 mm
	500 mm centres decrease	618 mm
	600 mm centres decrease	742 mm

Remember

The plumb and seat cuts for the jack rafter are the same as for the common rafter

K2. Know how to construct gables, verge and eaves

Constructing a roof would be relatively simple if it were only a matter of erecting and covering the rafters. However, life is never that simple. Roofs might need to accommodate all manner of additions that require a more flexible approach. A good carpenter or joiner will know how to construct the most common of these – gables, verges and eaves.

Gable ends

Gable ends are found on many houses and are the most common alternative to a hipped roof. The skills needed to construct a gable end are slightly different to those needed to build hipped roofs; the following section introduces those skills.

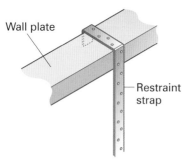

Figure 10.28 Restraint straps

Setting out for a gable end

First set out and fix the wall plate. The wall plate is set on the brick or block work and either bedded in by the bricklayer or temporarily fixed by nailing through the joints. Once secured it is held in place with restraint straps (see Figure 10.28). If the wall plate is to be joined in length, a **halving joint** is used (see Figure 10.29). It is vital that the wall plate is fixed level to avoid serious problems later.

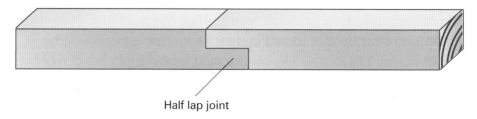

Figure 10.29 Plate with halving joint

Key terms

Halving joint – the same amount is removed from each piece of timber so that when fixed together the joint is the same thickness as the uncut timbers

Geometry – a form of mathematics using formulas, arithmetic and angles

Scale drawing – a drawing of a building, or component in the right proportions, but scaled down to fit on a piece of paper. On a drawing at a scale of 1:50, a line 10 mm long would represent 500 mm on the actual object

Sliding bevel – a tool that can be set so that the user can mark out any angle

Pattern rafter – a precisely cut rafter used as a template to mark out other rafters

Once the wall plate is in place, you need to measure the span and the rise. You can use these measurements to work out the rafter length in different ways, using a roofing square, a roofing ready reckoner, **geometry** or **scale drawings**. Ready reckoners and geometry are covered later in this unit, so we will start with scale drawings.

For this example we will use a span of 5 m and a rise of 2.3 m.

Using a scrap piece of plywood or hardboard, we first draw the roof to a scale that will fit the scrap piece of plywood/hardboard (usually a scale of 1:20).

From this drawing we can measure at scale and find the true length of the rafter. By using a **sliding bevel** we can work out the plumb cut and seat cut.

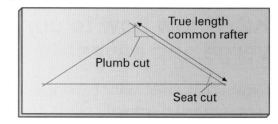

Figure 10.30 Sketch on a piece of scrap ply showing the true length, plumb and seat cut for a common rafter

Making and using a pattern rafter

From our scale drawing we can mark out one rafter, which we will then use as a **pattern rafter**.

There are five easy steps to follow when marking out a pattern rafter.

Figure 10.31 Step 1 Mark the pitch line

Figure 10.32 Step 2 Set the sliding bevel

Figure 10.33 Step 3 Mark the true length

Figure 10.34 Step 4 Mark out the seat cut

Step 1 Mark the pitch line one-third of the way up the width of the rafter.

Step 2 Set the sliding bevel to the plumb cut and mark the angle onto the top of the rafter.

Step 3 Mark the true length on the rafter, measuring along the pitch line.

Step 4 Use the sliding bevel to mark out the seat cut, then with a combination square mark out the birdsmouth at 90 degrees to the seat cut.

Figure 10.35 Step 5 Re-mark the plumb cut

Step 5 Re-mark the plumb cut to allow for half the thickness of the ridge.

Once it has all been marked out, this can be cut and used as a pattern rafter.

Did you know?

The first and last rafters are placed 50 mm away from the wall to prevent moisture that penetrates the outside wall coming into contact with the rafters, thus preventing rot

The pattern rafter can be used to mark out all the remaining common rafters, although it is advisable to mark out and cut only four, then place two at each end of the roof to check whether the roof is going to be level.

Once all the rafters are cut, mark out the wall plate and fix the rafters. Rafters are normally placed at 400 mm centres, with the first and last rafter 50 mm away from the gable wall. The rafters are usually fixed at the foot by skew-nailing into the wall plate and at the head by nailing through the ridge board.

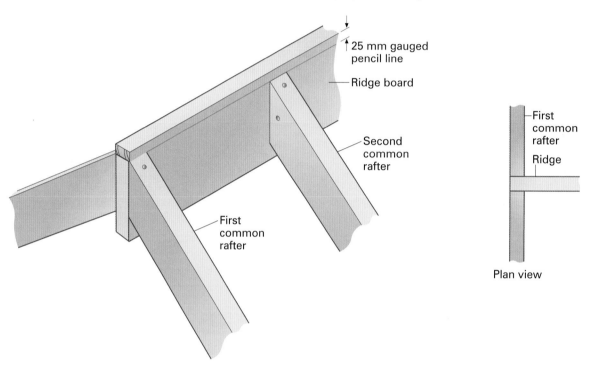

Figure 10.36 First two rafters fixed into place

If the roof requires a loft space, joists can be put in place and bolted to the rafters. If additional support is required, struts can be used.

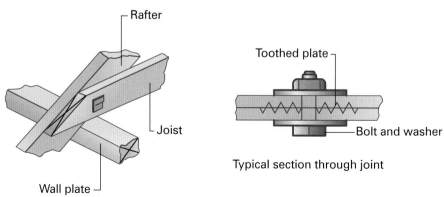

Figure 10.37 Rafter legs bolted to joists

For roofs with a large span, purlins provide adequate support. Purlins are usually built into the brickwork: either the gable wall will not have been finished being built yet, or the bricklayer will have left a **sand course**. The same is true of roof ladders (see below).

Finishing a gable end

There are two types of finish for a gable end:

- a flush finish – where the bargeboard is fixed directly onto the gable wall
- a roof ladder – a frame built to give an overhang and to which the bargeboard and soffit are fixed.

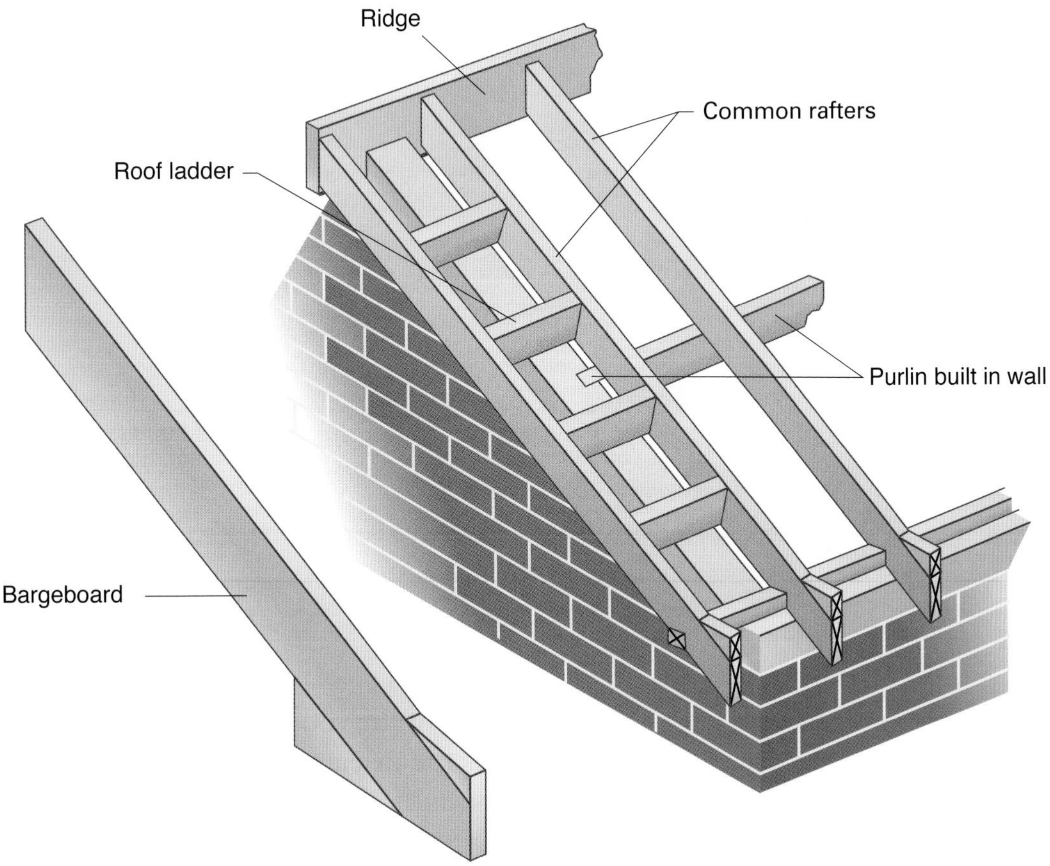

Key term

Sand course – where the bricklayer beds in certain bricks with sand instead of cement so that they can be easily removed to accommodate things like purlins

Figure 10.38 Roof ladder with bargeboard fitted

The most common way is to use a roof ladder, which when creating an overhang, prevents rainwater from running down the face of the gable wall.

The continuation of the fascia board around the verge of the roof is called the bargeboard. Usually the bargeboard is fixed to the roof ladder and has a built-up section at the bottom to encase the wall plate.

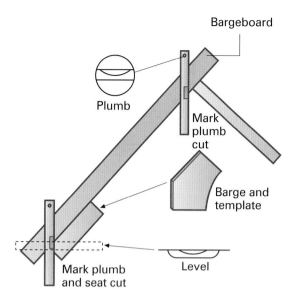

Figure 10.39 Marking out a bargeboard using a level

The simplest way of marking out the bargeboard is to temporarily fix it in place and use a level to mark the plumb and seat cut.

When fixing a bargeboard, the foot of the board may be mitred to the fascia, butted and finished flush with the fascia, or butted and extended slightly in front of the fascia to break the joint.

The bargeboard should be fixed using oval nails or lost heads at least 2.5 times the thickness of the board so that a strong fixing is obtained. If there is to be a joint along the length of the bargeboard, the joint *must* be a mitre.

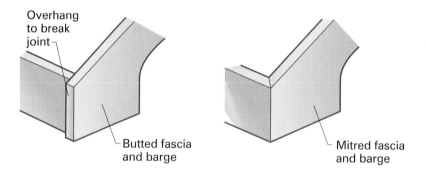

Figure 10.40 Fascia joined to bargeboard

Did you know?

Without soffit ventilation, air cannot flow through the roof space, which can cause problems such as dry rot

Eaves details

The eaves are how the lower part of the roof is finished where it meets the wall and incorporates the fascia and soffit. The fascia is the vertical board fixed to the ends of the rafters. It is used to close the eaves and allow fixing for rainwater pipes. The soffit is the horizontal board fixed to the bottom of the rafters and the wall. It is used to close the roof space to prevent birds or insects from nesting there and usually incorporates ventilation to help prevent rot.

There are various ways of finishing a roof at the eaves.

Flush eaves

Here the eaves are finished as close to the wall as possible. There is no soffit, but a small gap is left for ventilation.

Open eaves

Open eaves are where the bottom of the rafter feet are planed as they are exposed. The rafter feet project beyond the outer wall and eaves boards are fitted to the top of the rafters to hide the underside of the roof cladding. The rainwater pipes are fitted via brackets fixed to the rafter ends.

Closed eaves

Closed eaves are completely closed or boxed in. The ends of the rafters are cut to allow the fascia and soffit, to be fitted. The roof is ventilated either by ventilation strips incorporated into the soffit, or by holes drilled into the soffit with insect-proof netting over them. If closed eaves are to be re-clad due to rot, you must ensure that the ventilation areas are not covered up.

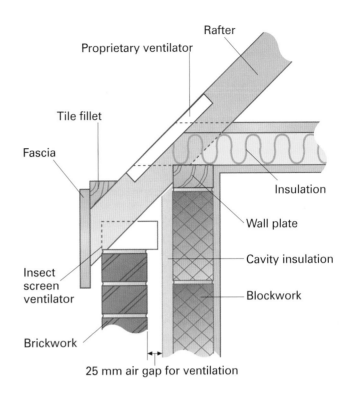

Figure 10.41 Flush eaves

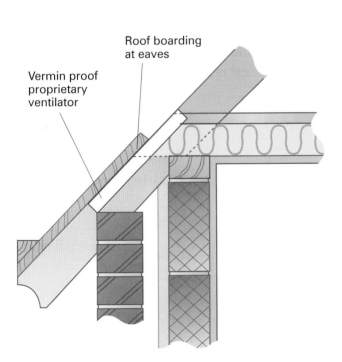

Figure 10.42 Open eaves

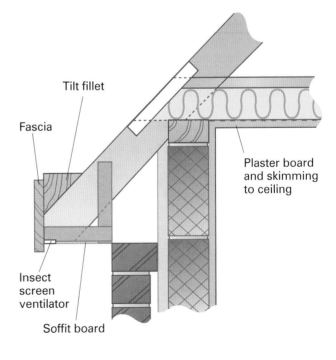

Figure 10.43 Closed eaves

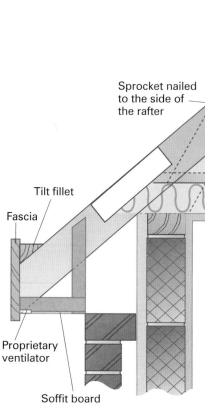

Sprocket nailed to the side of the rafter

Ceiling joists

Tilt fillet

Fascia

Proprietary ventilator

Soffit board

Figure 10.44 Sprocketed eaves

Sprocketed eaves

Sprocketed eaves are used where the roof has a sharp pitch. The sprocket reduces the pitch at the eaves, slowing down the flow of rainwater and stopping it overshooting the guttering. Sprockets can either be fixed to the top edge of the rafter or bolted onto the side.

Hipped roofs

In a fully hipped roof there are no gables and the eaves run around the perimeter, so there is no roof ladder or bargeboard.

Marking out for a hipped roof

All bevels or angles cut on a hipped roof are based on the right-angled triangle and the roof members can be set out using the following two methods:

- **roofing ready reckoner** – a book that lists in table form all the angles and lengths of the various rafters for any span or rise of roof (see pages 233–235)
- **geometry** – working with scale drawings and basic mathematic principles to give you the lengths and angles of all rafters.

Pythagoras' theorem

When setting out a hipped roof, you need to know Pythagoras' theorem. Pythagoras states that 'the square on the hypotenuse of a right-angled triangle is equal to the sum of the squares on the other two sides'. For the carpenter, the 'hypotenuse' is the rafter length, while the 'other two sides' are the run and the rise.

From Pythagoras' theorem, we get this calculation:

$A = \sqrt{B^2 + C^2}$ ($\sqrt{}$ means the square root and ^{2} means squared)

If we look at our right-angled triangle, we can break it down to:

A (the rafter length – the distance we want to know)
B^2 (the rise, multiplied by itself)
C^2 (the run, multiplied by itself)

Therefore, we have all we need to find out the length of our rafter (A):

$A = \sqrt{4^2 + 3^2}$
$A = \sqrt{4 \times 4 + 3 \times 3}$
$A = \sqrt{16 + 9}$
$A = \sqrt{25}$
$A = 5$

Thus, our rafter would be 5 m long.

Functional skills

To find true lengths you will need to use the correct mathematical procedure. This will allow you to practice FM 1.1.1 identify and select mathematical procedures, FM 1.2.1b interpret information from sources such as diagrams, tables, charts and graphs and FM 1.2.1c draw shapes. Practice in the skill of problem solving is also covered in these sections.

Did you know?

The three angles in a triangle always add up to 180 degrees

Finding true lengths

The next task is to find the hip rafter true length, plumb and seat cuts. This is done in two stages (the first will be familiar to you, as it is the same as for a common rafter).

The next step is to lengthen the common rafter true length by the amount of the rise, then join this line up to the base of the roof.

From this point, to make the geometrical drawings as clear as possible, abbreviated labels will be used. See Table 10.1.

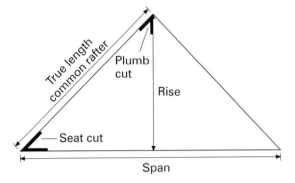

Figure 10.45 Finding true length common rafter

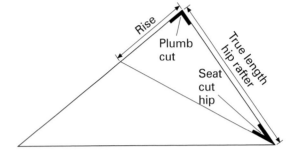

Figure 10.46 Finding true length hip rafter

Table 10.1 Abbreviations

Abbreviation	Definition
TL	true length
TLCR	true length common rafter
TLHR	true length hip rafter
TLJR	true length jack rafter
TLCrR	true length cripple rafter
PC	plumb cut
SC	seat cut
PCCR	plumb cut common rafter
SCCR	seat cut common rafter
PCHR	plumb cut hip rafter
SCHR	seat cut hip rafter
PCVR	plumb cut valley rafter
SCVR	seat cut valley rafter
EC	edge cut
ECHR	edge cut hip rafter
ECJR	edge cut jack rafter
ECVR	edge cut valley rafter
ECCrR	edge cut cripple rafter

Did you know?

Most drawings use abbreviations or symbols to avoid cluttering the drawings and make them easier to read

There are two other angles that are concerned with hip rafters: the dihedral angle (or backing bevel for the hip) and the edge cut to the hip.

Finding the dihedral or backing bevel angle

The backing bevel angle is the angle between the two sloping roof surfaces. It provides a level surface so that the tile battens, or roof boards, can lie flat over the hip rafters. The backing bevel angle is rarely used in roofing today as the edge of the hip is usually worked square, but you should still know how to work it out.

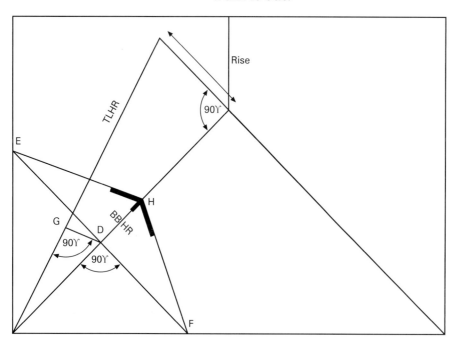

Figure 10.47 Dihedral angle for hip

Step 1 Draw a plan of the roof and mark on the TLHR as before.

Step 2 Draw a line at right angles to the hip on the plan at D, to touch the wall plates at E and F.

Step 3 Draw a line at right angles to the TLHR at G, to touch point D.

Step 4 With centre D and radius DG, draw an arc to touch the hip at H.

Step 5 Join E to H and H to F. This gives the required backing bevel (BBHR).

Finding the edge cut

The edge cut is applied to both sides of the hip rafter at the plumb cut. It enables the hip to fit up to the ridge board between the crown and the common rafters.

Step 1 Draw a plan of the roof and mark on the TLHR as before.

Step 2 With centre I and radius IB, swing the TLHR down to J, making IJ, TLHR.

Step 3 Draw lines at right angles from the ends of the hips and extend the ridge line. All three lines will intersect at K.

Step 4 Join K to J. Angle IJK is the required edge cut (ECHR).

Unit 2010 How to erect structural carcassing

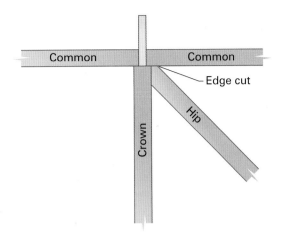

Figure 10.48 Edge cut on hip joined to ridge

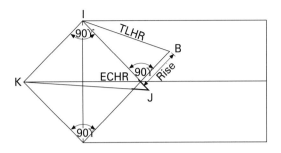

Figure 10.49 Edge cut on hip

The jack rafter's plumb and seat cuts are the same as those used in the common rafters, so all you need to work out is the true length and edge cut.

Step 1 Draw the plan and section of the roof. Mark the jack rafters on the plan. Develop roof surfaces by swinging TLCR down to L and projecting down to M^1.

Step 2 With centre N and radius NM^1, draw arc M^{1}O. Join points M^1 and O to ends of hips as shown.

Step 3 Continue jack rafters on to development.

Step 4 Mark the true length of jack rafter (TLJR) and edge cut for jack rafter (ECJR).

Setting out a hipped end

First, we need to fit the wall plate, then the ridge and common rafters. To know where to place the common rafters you need to work out the true length of the rafter, then begin to mark out the wall plate.

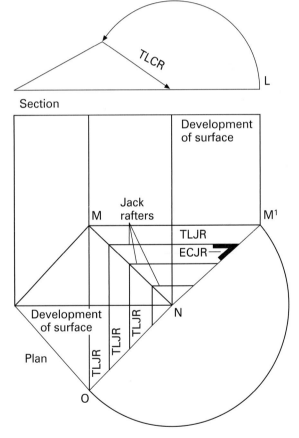

Figure 10.50 Jack rafter length and cuts

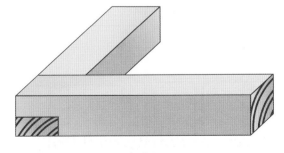

Figure 10.51 Corner halving

The wall plate is joined at the corners and marked out as shown in Figure 10.52.

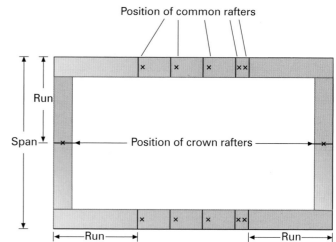

Figure 10.52 Wall plate marked out

Marking out for a hipped roof

To mark out for a hipped roof, follow these steps:

Step 1 Measure the span and divide it by two to get the run, then mark this on the hipped ends. The centre of your **crown rafter** will line up with this mark.

Step 2 Mark the run along the two longer wall plates. These marks will give you the position of your first and last common rafter.

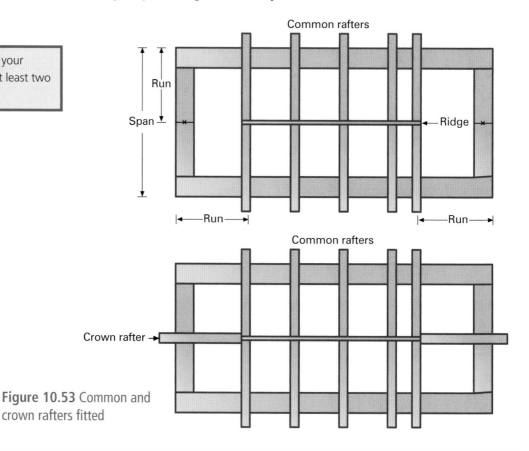

Figure 10.53 Common and crown rafters fitted

Step 3 The common rafter will sit to the side of this line, so a cross or other mark should be made to let you know on which side of the line the rafter will sit.

Step 4 Mark positions for the rest of the rafters on the wall plate at the required centres, again using a cross or other mark to show you on which side of the line the rafter will sit. Note: the last two rafters may be closer together than the required centres, but must not be wider apart than the required centres.

Step 5 Cut and fit the common rafters using the same method as used for a gable roof.

Step 6 Fit the crown rafter, which has the same plumb and seat cuts as the common rafter and is almost the same length, but here you should *not* remove half the ridge thickness.

Step 7 For the hip rafters, work out the true length, all the angles and bevels and mark out one hip. Then cut the hip and try it in the four corners. If the hip fits in all four corners you can use it as a template to mark out the rest of the hips. If not, the roof is out of square or level, but you can still use this hip to help mark out the remaining three corners.

With a hip rafter it is important to remember that the pitch line is marked out differently. It is marked from the top of the rafter and is set at two-thirds of the depth of the common rafter.

The best way to check your rafter before cutting is to measure from the point of the ridge down to the corner of the wall plate. This distance should be the same as marked on your rafter.

Step 8 Cut in the jack rafters. Find out the true length and edge cut of all the jack rafters, then mark them out and cut them. As the jack rafters are of different sizes it is better to cut them individually to fit. They can still be used as template rafters on the opposite side of the hip.

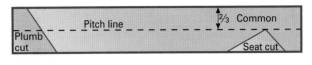

Figure 10.80 Pitch line marked on hip

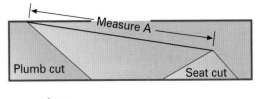

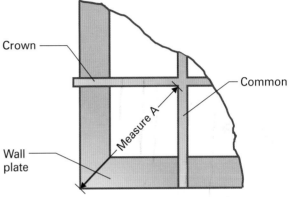

Figure 10.54 Checking hip measurements

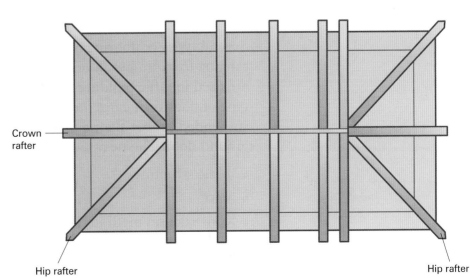

Figure 10.55 Hip rafters in place

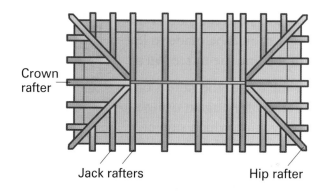
Figure 10.56 Jacks fitted

Crown rafter

Jack rafters

Hip rafter

Working life

Chloe and Tyrone have been tasked with cutting in a garage roof with gable ends. They work out the angles and lengths, cut the pattern rafter, which they use to mark out all the others, then fit them. They are almost finished when the foreman turns up and tells them to stop and check it for level. Chloe and Tyrone put a level on the ridge board and find that it is way out of level.

How could this have happened? What could have been done to prevent it?

K3. Know how to install floor joists

Without adequate support, buildings would collapse in on themselves. All of the different structural elements play a role in strengthening the building and floor joists are no different. Roofs, walls and floors all work together to keep the building upright and solid.

Ground and upper floors

Several types of flooring are used in the construction of buildings, ranging from timber floors to large pre-cast concrete floors, which are used in large buildings such as residential flats. The main type of flooring a carpenter will deal with is suspended timber. A suspended timber floor can be fitted at any level from top floor to ground floor. The next few pages will look at:

- basic structure
- joists
- construction methods
- floor coverings.

Basic structure

Suspended timber floors are constructed with timbers known as joists, which are spaced parallel to each other spanning the distance of the building. Suspended timber floors are similar to traditional roofs in that they can be single or double, a single floor being supported at the two ends only and a double floor supported at the two ends and in the middle by way of a sleeper/dwarf wall, steel beam or load-bearing partition.

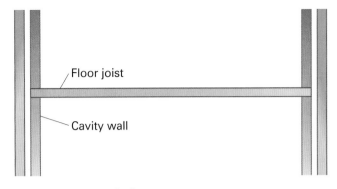

Figure 10.57 Single floor

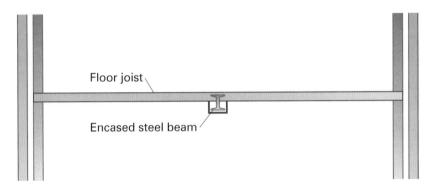

Figure 10.58 Double floor

All floors must be constructed to comply with the *Building Regulations*, in particular Part C, which is concerned with damp. The bricklayer must insert a **damp proof course (DPC)** between the brick or block work when building the walls, situated no less than 150 mm above ground level. This prevents moisture moving from the ground to the upper side of the floor. No timbers are allowed below the DPC. Air bricks, which are built into the external walls of the building, allow air to circulate round the underfloor area, keeping the moisture content below the dry rot limit of 20 per cent, thus preventing dry rot.

Key term

Damp proof course (DPC) – a substance that is used to prevent damp from penetrating a building

Joists

In domestic dwellings, suspended upper floors are usually single floors, with the joist supported at each end by the structural walls but, if support is required, a load-bearing partition is used. The joists that span from one side of the building to the other are called bridging joists, but any joists that are affected by an opening in the floor such as a stairwell or chimney are called trimmer, trimming and trimmed joists.

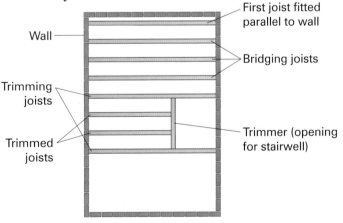

Figure 10.59 Joists and trimmers

Bridging joists are usually sawn timber 50 mm thick. The trimmer that carries the trimmed joists and transfers this load to the trimming must be thicker – usually 75 mm sawn timber, or in some instances two 50 mm bridging joists bolted together. The depth of the joist is easily worked out by using the calculation:

Depth of joist = span / 20 + 20

So, for example, if you have a span of 4 m

Depth = 4000 / 20 + 20
Depth = 200 – 20
Depth = 220

The depth of the joist required would be 220 mm.

If the span was 8 m, the depth would double to 440 mm. A depth of 440 mm is too great, so you would need to look at putting in a support to create a double floor.

Timbers used in carcassing work such as joists or rafters, must be treated with preservative to protect them from rot and insect attack. It is also important to ensure that there is good ventilation within any roof or floor space, as this will prevent the moisture content in the timbers from reaching a high level. Moisture in timber can lead to rot.

Types of joist

As well as the traditional method of using solid timber joists, there are now alternatives available. These are the most common.

Laminated joists

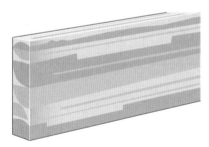

Figure 10.60 Laminated joist

These were originally used for spanning large distances, as a laminated beam could be made to any size, but now they are more commonly used as an environmental alternative to solid timber. Recycled timber can be used in the laminating process. They are more expensive than solid timber as the joists have to be manufactured.

I type joists

Figure 10.61 I type joist

These are some of the most commonly used joists in the construction industry. They are particularly popular in new build and are the only joists used in timber kit house construction. I type beam joists are lighter and more environmentally friendly, as they use a composite panel in the centre, usually made from oriented strand board (OSB), which can be made from recycled timber.

The following construction methods show how to fit solid timber joists, but whichever joists you use, the methods are the same.

Construction methods

A suspended timber floor must be supported either end. Figures 10.62 and 10.63 show ways of doing this.

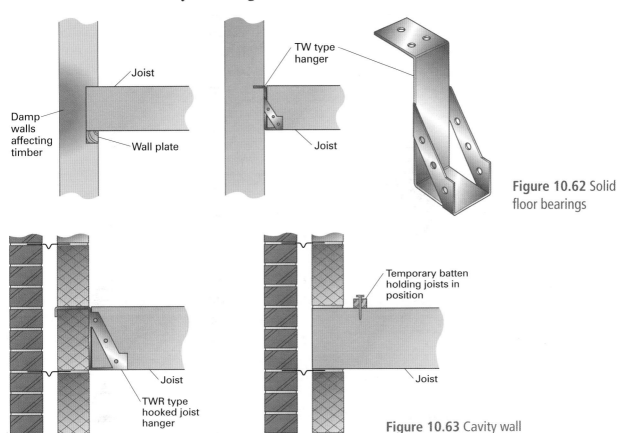

Figure 10.62 Solid floor bearings

Figure 10.63 Cavity wall bearings

If a timber floor has to trim an opening, there must be a joint between the trimming and the trimmer joists. Traditionally, a **tusk tenon joint** was used (even now, this is sometimes preferred) between the trimming and the trimmer joist. If the joint is formed correctly, a tusk tenon is extremely strong, but making one is time-consuming. A more modern method is to use a metal framing anchor or timber-to-timber joist hanger.

Key term

Tusk tenon joint – a type of mortise and tenon joint that uses a wedge-shaped key to hold the joint together

Figure 10.64 Traditional tusk tenon joint

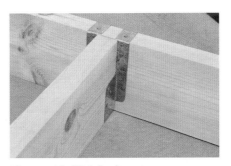

Figure 10.65 Joist hanger

Fitting floor joists

Before the carpenter can begin constructing the floor, the bricklayer needs to build the honeycomb sleeper walls. This type of walling has gaps in each course to allow the free flow of air through the underfloor area. It is on these sleeper walls that the carpenter lays his timber wall plate, which will provide a fixing for the floor joists.

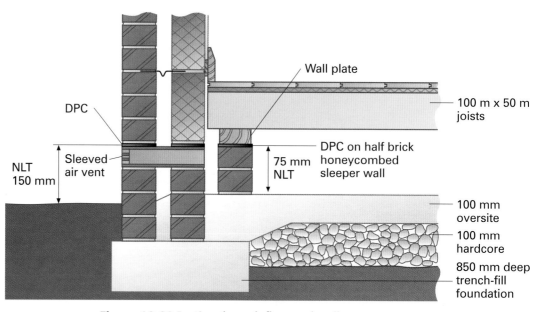

Figure 10.66 Section through floor and wall

The following pages describe the steps in fitting floor joists.

Step 1 Bed and level the wall plate onto the sleeper wall with the DPC under it.

Step 2 Cut joists to length and seal the ends with a coloured preservative. Mark out the wall plate with the required centres, space the joists out and fix temporary battens near each end to hold the joists in position. Ends should be kept away from walls by approximately 12 mm. It is important to ensure that the camber is turned upwards.

Step 3 Fix the first joist parallel to the wall with a gap of 50 mm. Fix trimming and trimmer joists next to maintain the accuracy of the opening.

Step 4 Fix subsequent joists at the required spacing as far as the opposite wall. Spacing will depend on the size of joist and/or floor covering, but 400 mm to 600 mm centres are usually used.

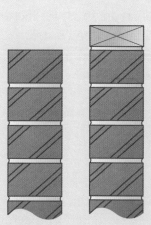

Figure 10.67 Step 1 Bed in the wall plate

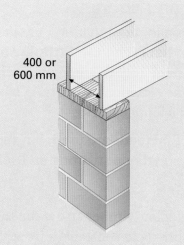

400 or 600 mm

Figure 10.68 Step 2 Space out joists

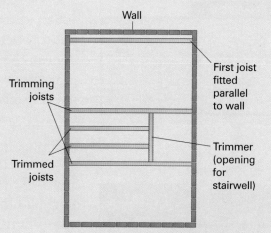

Wall

Trimming joists

First joist fitted parallel to wall

Trimmer (opening for stairwell)

Trimmed joists

Figure 10.69 Step 3 Fit first joist and trimmers

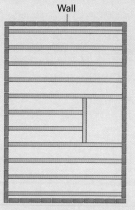

Wall

Figure 10.70 Step 4 Fit remaining joists

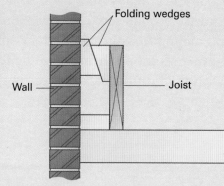

Folding wedges

Wall

Joist

Figure 10.71 Step 5 Fit folding wedges

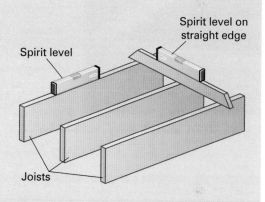

Spirit level on straight edge

Spirit level

Joists

Figure 10.72 Step 6 Ensure joists are level

Step 5 Fit folding wedges to keep the end joists parallel to the wall. Overtightening is to be avoided in case the wall is strained.

Step 6 Check that the joists are level with a straight edge or line and, if necessary, pack with slate or DPC.

Step 7 Fit restraining straps and, if the joists span more than 3.5 m, fit strutting and bridging, described in more detail next.

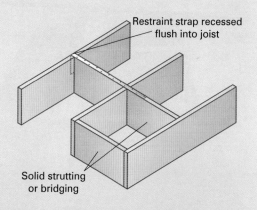

Restraint strap recessed flush into joist

Solid strutting or bridging

Figure 10.73 Step 7 Fix restraining straps, struts and bridges

Remember

It is very important to clean the underfloor area before fitting the flooring, as timber cuttings or shavings are likely to attract moisture

Strutting and bridging

When joists span more than 3.5 m, a row of struts must be fixed midway between each joist. Strutting or bridging stiffens the floor in the same way that noggins stiffen timber stud partitions, preventing movement and twisting, which is useful when fitting flooring and ceiling covering.

A number of methods are used, but the main ones are solid bridging, herringbone strutting and steel strutting.

Solid bridging

For solid bridging, timber struts the same depth as the joists are cut to fit tightly between each joist and are skew-nailed in place. A disadvantage of solid bridging is that it tends to loosen when the joists shrink.

Herringbone strutting

Here timber battens (usually 50 × 25 mm) are cut to fit diagonally between the joists. A small saw cut is put into the ends of the battens before nailing to avoid the battens splitting. This will remain tight even after joist shrinkage. The following steps describe the fitting of timber herringbone strutting.

Figure 10.74 Solid bridging

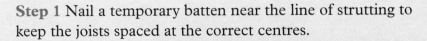

Step 1 Nail a temporary batten near the line of strutting to keep the joists spaced at the correct centres.

Figure 10.75 Step 1 Space joists

Figure 10.76 Step 2 Mark joist depths

Figure 10.77 Step 3 Lay struts across two joists at a diagonal

Figure 10.78 Step 4 Cut to the mark

Step 2 Mark the depth of a joist across the edge of the two joists, then measure 12 mm inside one of the lines and remark the joists. The 12 mm less than the depth of the joist ensures that the struts will finish just below the floor and ceiling level (as shown in Step 5).

Step 3 Lay the strut across two joists at a diagonal to the lines drawn in Step 2.

Step 4 Draw a pencil line underneath as shown in Step 3 and cut to the mark. This will provide the correct angle for nailing.

Step 5 Fix the strut between the two joists. The struts should finish just below the floor and ceiling level. This prevents the struts from interfering with the floor and ceiling if movement occurs.

Figure 10.79 Step 5 Fix the strut

Steel strutting

There are two types of galvanised steel herringbone struts available.

The first has angled lugs for fixing with the minimum 38 mm round head wire nails.

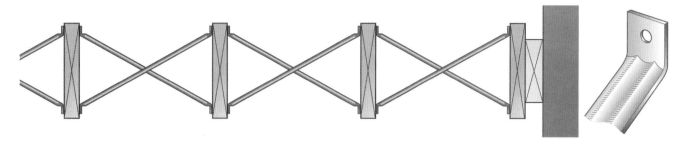

Figure 10.80 Catric® steel joist struts

The second has pointed ends, which bed themselves into joists when forced in at the bottom and pulled down at the top. Unlike other types of strutting, this type is best fixed from below.

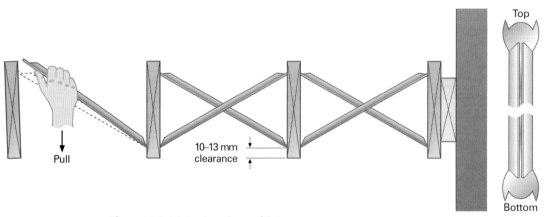

Figure 10.81 Batjam® steel joist struts

The disadvantage of steel strutting is that it only comes in set sizes to fit centres of 400, 450 and 600 mm. This is a disadvantage as there will always be a space in the construction of a floor that is smaller than the required centres.

Restraint straps

Anchoring straps, normally referred to as restraint straps, are needed to restrict any possible movement of the floor and walls due to wind pressure. They are made from galvanised steel, are 5 mm thick for horizontal restraints and 2.5 mm for vertical restraints, 30 mm wide and up to 1.2 m in length. Holes are punched along the length to provide fixing points.

When the joists run parallel to the walls, the straps will need to be housed into the joist to allow the strap to sit flush with the top of the joist, keeping the floor even. The anchors should be fixed at a maximum of 2 m centre to centre. More information can be found in schedule 7 of the *Building Regulations*.

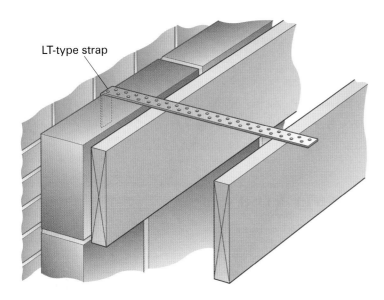

LT-type strap

Figure 10.82 Restraint straps for joists parallel or at right angles to a wall

FAQ

Why are flat roofs only guaranteed for a certain amount of time?

Most things that you buy, or have fitted, have a guarantee for a certain amount of time and building work is no different. If the flat roof had a guarantee for 50 years, the builder would be responsible for any maintenance work on the roof free of charge for 50 years. Since the average life of a flat roof is 12–15 years, the builder will only offer a guarantee for 10 years.

Which type of flat roof decking is the best?

There is no best or worst but some materials are better than others. All the materials stated serve a purpose, but only if they are finished correctly. For example, a chipboard-covered roof that is poorly felted will leak, as will a metal roof if the screw or bolt holes are not sealed correctly.

How can I get the lead on a flat roof to fit into the brick or block work?

The lead is fitted into a channel or groove that is cut into the brick or block work by a bricklayer. The groove is cut using either a disc cutter or a hammer and cold chisel. Once the groove is cut, the lead is fed into the groove, wedged and sealed with a suitable mastic or silicon.

Check it out

1 Describe the difference between a single and double roof.
2 Explain the reason for having a mansard roof.
3 Explain the difference between a hip and a gable end.
4 What is a pitch line and where is it marked?
5 State a suitable way of fitting a wall plate.
6 State the distance which the first and last common rafters must be away from the gable wall.
7 What is the dihedral angle (backing bevel) and what purpose does it serve?
8 Explain the purpose of a roof ladder.
9 Explain what type and size of nail should be used to fix bargeboard.
10 Explain why it important that a carpenter learns Pythagoras' theorem.
11 Why do you not need to work out the plumb and seat cuts for the jack or cripple rafters?
12 Describe two different ways of forming a valley.
13 Explain why you need to add extra rafters when trimming an opening.
14 State two different ways of providing a fall on a flat roof.
15 Name four different materials used when decking a flat roof.
16 Explain the purpose of strutting a floor or flat roof.
17 What is the purpose of access hatches or traps in floors?

Getting ready for assessment

The information contained in this unit, as well as continued practical assignments that you will carry out in your college or training centre, will help you with preparing for both your end of unit test and the diploma multiple-choice test. It will also aid you in preparing for the work that is required for the synoptic practical assignments.

The information contained within this unit will aid you in learning how to identify and calculate the materials and equipment required for erecting structural carcassing.

You will need to be familiar with:

- preparing materials for carrying out a range of carcassing tasks
- erecting truss rafter roofs
- constructing gables, verges and eaves
- installing floor joists.

This unit will have made you familiar with the methods of erecting structural carcassing. For example, learning outcome two has shown you the methods of finishing gables and eaves and the tools that are best used for these jobs. You will need to use the knowledge and the specifications to construct and fix these parts of a building. You will need to remember the safety issues for working at height from Unit 1001 on pages 26–33 when working on roofs. Safety is a major issue when working at height, and you will need to be sure that you are avoiding the common hazards associated with this type of work.

You will also need to use your knowledge of materials to select the best suited for a particular job. To do this you will need to study the specifications and take into account the particular location you will be building in.

Before you start work on the synoptic practical test it is important that you have had sufficient practice and that you feel that you are capable of passing. It is best to have a plan of action and a work method that will help you. You will also need a copy of the required standards, any associated drawings and sufficient tools and materials. It is also wise to check your work at regular intervals. This will help you to be sure that you are working correctly and help you to avoid problems developing as you work.

Your speed at carrying out these tasks will also help you to prepare for the time limit that the synoptic practical task has. But remember, don't try to rush the job as speed will come with practice and it is important that you get the quality of workmanship right.

Always make sure that you are working safely throughout the test. Make sure you are working to all the safety requirements given throughout the test and wear all appropriate personal protective equipment. When using tools, make sure you are using them correctly and safely.

Good luck!

CHECK YOUR KNOWLEDGE

1 A roof with a pitch of 10° or less is called a:
  a lean to.
  b pitch roof.
  c flat roof.
  d mono pitch.

2 The element of a roof that acts as a spine at the apex is the:
  a ridge board.
  b purlin.
  c hip.
  d gable.

3 Wall plates are used to:
  a allow the guttering to be fixed.
  b allow the rafters to be fixed and spread the load.
  c allow the fascia to be fixed.
  d keep the rafters straight.

4 The diminishing rafters that are fixed to the hip are:
  a crown rafters.
  b common rafters.
  c jack rafters.
  d cripple rafters.

5 The distance measured between two wall plates is called the:
  a rise.
  b span.
  c run.
  d pitch.

6 The pitch of a roof is measured in:
  a centimetres.
  b degrees.
  c metres.
  d ratios.

7 The angles and lengths of a pitch roof can be worked out by using:
  a a ready reckoner.
  b a roofing square.
  c geometry.
  d any of the above.

8 A floor that is supported in the middle is known as a:
  a single floor.
  b double floor.
  c triple floor.
  d suspended floor.

9 The joist that spans the entire floor is called a:
  a trimmer joist.
  b trimmed joist.
  c trimming joist.
  d bridging joist.

10 The best way to join a trimmer to a trimming joist is to use:
  a 5″ nails.
  b 5″ screws.
  c metal timber connectors.
  d a halving joint.

11 The first and last joist must be what distance from the wall?
  a 40 mm.
  b 50 mm.
  c 60 mm.
  d 75 mm.

12 When fixing floor boards, the nails should be:
  a 2 x the thickness of the board.
  b 2 ½ x the thickness of the board.
  c 3 x the thickness of the board.
  d 3 ½ x the thickness of the board.

13 The joints between floorboards should be glued and screwed to prevent the floor:
  a squeaking.
  b rotting.
  c leaking.
  d collapsing.

UNIT 2011

How to carry out maintenance

In time, all components within a building will deteriorate. Timber will rot, brickwork will crumble and even the ironmongery fail through wear and tear. These components need to be maintained or repaired to prevent the building falling into a state of disrepair. General building maintenance is now a recognised qualification, but traditionally this work is done by carpenters as they have the best links to other trades and the best understanding of what they do.

This unit also supports NVQ Unit VR12 Maintain Non-structural Carpentry Work.

This unit contains material that supports TAP Unit 9 Maintain Internal and External Timber Components. It also contains material that supports the delivery of the five generic units.

This unit will cover the following learning outcomes:

- How to repair mouldings, doors and windows
- How to replace gutters and down pipes
- How to replace sash cords
- How to make good plaster, paintwork and brickwork.

Unit 2011

How to carry out maintenance

Functional skills

When reading and understanding the text in this unit, you are practicing several functional skills.

FE 1.2.1 – Identifying how the main points and ideas are organised in different texts.

FE 1.2.2 – Understanding different texts in detail.

FE 1.2.3 – Read different texts and take appropriate action, e.g. respond to advice/instructions. If there are any words or phrases you do not understand, use a dictionary, look them up using the internet or discuss with your tutor. FM 1.2.1b relates to interpreting information from sources such as diagrams, tables, charts and graphs

K1. Know how to repair mouldings, doors and windows

While carpenters and joiners will spend a lot of time making things, repair skills are as important as construction skills. No one would buy a new house because their door was binding, but they are very likely to call you in to fix it! Knowing how to repair key items like decorative mouldings, doors and windows is vital to your success in the industry. There are other repair jobs, not related to carpentry, that you might also be called upon to do – making good minor damage to brickwork or plaster, for example. In this unit you will learn the skills needed.

Timber decay

Repairs to windowsills or frames are often needed as a result of timber decay. You will need to know how to identify the different types of timber decay and the preservatives you can use to prevent future decay. For an in-depth look at wet and dry rot, a timber identification table and a reminder of some common timber defects, see Unit 2008, pages 135–139.

The decay of timber is caused by one or both of the following:

- wood-destroying fungi (covered on pages 137–139)
- wood-boring insects.

Wood-boring insects

Many insects attack or eat wood, causing structural damage. There are four main types of insects found in the UK, all classed as beetles.

Common furniture beetle

This wood-boring insect can damage both softwoods and hardwoods. The larvae of the beetle bore through the wood, digesting the cellulose. After about three years, the beetle forms a pupal chamber near the surface, where it changes into an adult beetle. In the summer, the beetle bites its way out to the surface, forming the characteristic round flight hole, measuring about 1.5 mm in diameter. After mating, the females lay their eggs (up to 80) in cracks, crevices or old flight-holes. The eggs hatch and a new generation begins a fresh life cycle.

Death-watch beetle

This wood-boring insect is related to the common furniture beetle, but is much larger, with a flight-hole of about 3 mm in diameter, usually found in decaying oak. The female lays up to 200 eggs.

While generally attacking hardwoods only, this wood-boring insect has been known to feed on decaying softwood timbers. The Death Watch Beetle is well-known for making a tapping sound, caused by the head of the male during the mating season.

Powder post beetle

This beetle gets its name from the way it can reduce timber to a fine powder. Powder post beetles generally attack timber with a high moisture content. As with all other beetles, the female lays eggs and the larvae do the damage.

House longhorn beetle

This wood-destroying insect attacks seasoned softwoods, laying its eggs in the cracks and crevices of wood. In Great Britain, this insect is found mainly in Surrey and Hampshire.

As well as beetles, weevils and wood wasps can also cause problems.

Wood-boring weevil

This is a wood-boring insect similar in appearance and size to the common furniture beetle. It differs in that it will only attack timber that is already decayed by wood-rotting fungi. There are

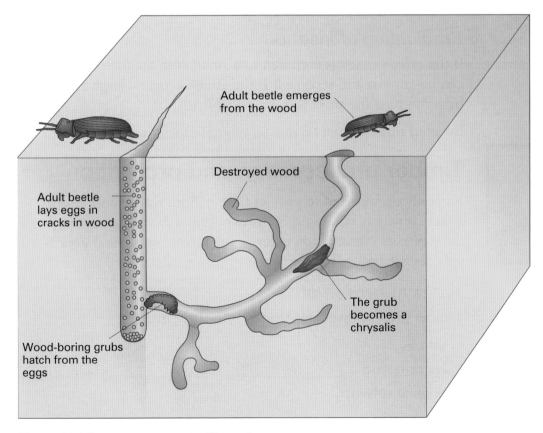

Adult beetle emerges from the wood

Destroyed wood

Adult beetle lays eggs in cracks in wood

The grub becomes a chrysalis

Wood-boring grubs hatch from the eggs

Figure 11.1 Insects can cause significant damage

over 50,000 species of weevil, all of which have long snouts. The wood-boring weevil is prolific, and is known to have up to two complete life cycles in one year. Its presence may therefore be accompanied by serious structural collapse of timber.

Wood wasp

These insects are also known as horntails. They are wasp-like insects and may be seen or heard buzzing and making cosmetic damage around the home. The body is 2–5 cm long and is multi-coloured, blue or black, with wings of matching colour. Walls, wood floors, doors and other wooden surfaces may have holes that are approximately 3 mm in diameter. These holes may also appear on items covering wooden structures such as carpeting, wallpaper, and linoleum. Wood wasps lay their eggs in forest trees that are either damaged or weak. It can take up to five years for adult wood wasps to emerge, so they are often found in trees that have been felled for construction purposes – often even after the wood has been stored. However, most wood wasp holes will appear within the first two years after use of the cut wood. The good news is that these insects only lay their eggs in weakened forest trees and will not re-infest buildings or household structures.

Eradication of insects

If the insect attack is confined to a small area the infected timbers can be cut out and replaced. If the attack is over a large area it is better to pass the work to a specialist firm.

Infected timber must be moved to a safe location and burned.

Timber preservation and protection

Not only do we need to protect timber from fungal and insect attack, but also, where timber is exposed, it needs protecting from the weather. Preservation extends the life of timber and greatly reduces the cost of maintenance.

Types of preservative

Timber preservatives are divided into three groups:

- tar oils
- water-borne
- organic solvents.

Remember

Contact a professional for advice where load-bearing timbers have been attacked, as there may be a need to prop or shore up to prevent collapse of the structure

Tar oils

Tar oils, derived from coal tar, are very effective and relatively cheap. However, they give off a very strong odour which may contaminate other materials. Creosote is the most common type of this preservative.

Water-borne

Water-borne preservatives are mainly solutions of copper, zinc, mercury or chrome. Water is used to carry the chemical into the timber and then allowed to evaporate, leaving the chemical in the timber. They are very effective against fungi and insects and are able to penetrate into the timber. They are also easily painted over and relatively inexpensive.

Organic solvents

Organic solvents are the most effective, but also the most costly of the preservatives. They have excellent penetrating qualities and dry out rapidly. Many of this type are proprietary brands such as those manufactured by Cuprinol™.

Methods of application

Preservatives can be applied in two ways:

- non-pressure methods – brushing, spraying, dipping or steeping
- pressure methods – empty cell, full cell or double vacuum.

Non-pressure methods

Although satisfactory results can be achieved, there are disadvantages with using non-pressure methods. The depth of penetration is uneven and, with certain timbers, impregnation is insufficient to prevent leaking out (leaching). Table 11.1 lists non-pressure methods and how and where to employ them.

Table 11.1 Non-pressure methods

Method	How and where to employ it
Brushing	The most commonly used method of applying preservative, it is important to apply the preservative liberally and allow it to soak in.
Spraying	Usually used where brushing is difficult to carry out, in areas such as roof spaces.
Dipping	Timbers are submerged in a bath of preservative for up to 15 minutes.
Steeping	Similar to dipping, only the timber is left submerged for up to two weeks.

Pressure methods

Pressure methods generally give better results with deeper penetration and less leaching. Table 11.2 lists pressure methods and how and where to employ them.

Table **11.2** Pressure methods

Method	How and where to employ it
Empty cell	Preservative is forced into the timber under pressure. When the pressure is released the air within the cells expands and blows out the surplus for re-use. This method is suitable for water-borne and organic solvent preservatives.
Full cell	Similar to empty cell, but prior to impregnation a vacuum is applied to the timber. The preservative is then introduced under pressure to fill the cells completely. Suitable for tar oils and water-borne preservatives.
Double vacuum	A vacuum is applied to remove air from the cells, the preservative is introduced, the vacuum is released and pressure is applied. The pressure is released and a second vacuum applied to recover surplus preservative. This method is used for organic solvent preservatives.

Safety tip

All preservatives are toxic and care should be taken at all times. Protective clothing should always be worn when using preservatives

Using preservatives: Structural timbers

The maintenance of structural timbers is vital as they will almost certainly be carrying a load; joists carry the floors above, while rafters carry the weight of the roof. Because of this, structural repairs should be carried out by qualified specialists, so this section will give a brief overview of the work that is involved.

Joists

Joist ends are susceptible to rot: they are close to the exterior walls and can be affected in areas like bathrooms if the floor gets soaked and does not dry out. Joists can also be attacked by wood-boring insects. For both rot and insect damage, the repair method is the same.

Shore the area, with the weight spread over the props, then lift the floorboards to see what the problem is (for the purposes of this example, dry rot) and throw the old floorboards away.

Before making any repairs, you need to find and fix the cause of the problem, otherwise the same problems will keep arising. Rot at the ends of joists is usually caused by poor ventilation, for example the cavity or the airbricks may be blocked.

Next cut away the joists allowing 600 mm into sound timber and treat the cut ends with a suitable preservative.

New pre-treated timbers should be laid and bolted onto the existing joists with at least 1 m overlap. New timbers should ideally be placed either side of the existing joist and in place of the removed timber.

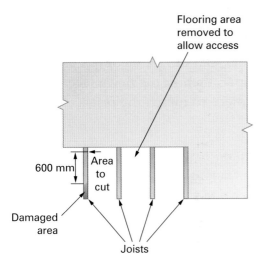

Figure 11.2 Cut away the rotten area of the joist

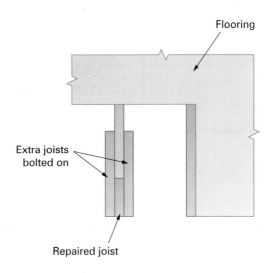

Figure 11.3 Repaired joist

Any new, untreated timbers should be treated with a suitable preservative, before or after laying. Finally, slowly remove the props and make good as necessary.

Rafters

Rafters are susceptible to the same problems as joists and especially to rot caused by a lack of ventilation in the roof space. Again the problem needs to be remedied before any repairs start.

Rafters are replaced or repaired in the same way as joists. They are easier to access, but shoring rafters can be difficult. In some cases it is best to strip the tiles and felt from the affected portion of the roof before starting.

For trussed roofs you should contact the manufacturer. Trusses are stress graded to carry a certain weight, so attempting to modify them without expert advice could be disastrous.

Repairing and replacing architrave and mouldings

Mouldings such as skirting or architrave, rarely need repairing, usually only because of damp or damage when moving furniture, and in most cases it is easier to replace them than repair them.

Safety tip

When repairing or replacing joists, you must first shore up the ceiling to carry the weight while you work. Do not alter or remove the shoring until the job is complete

Architrave

To replace architrave, you simply remove the damaged piece and fit a new piece.

First, check that the piece you are removing is not nailed to an existing piece. Then run a sharp utility knife down both sides of the architrave, so that when it is removed it does not damage the surrounding decorations or remaining architrave.

Once the old piece is removed, clean the frame of old paint, give it a light sanding with sandpaper, then fit the new piece. Finally, paint, stain or finish the new piece to match the rest.

Skirting board

Replacing skirting boards is slightly more difficult than replacing architrave. The way skirting boards are fitted could mean that the board you wish to replace is held in place by other skirting boards. Rather than remove the other skirting boards, you should cut or drill a series of holes in the middle of the board, splitting it in two so that you can remove it that way. Again, running a utility knife along the top of the skirting board will avoid unnecessary damage to surrounding decorations. Once the old skirting has been removed, the new piece can be fitted and finished to match the existing skirting.

You can replace other mouldings such as picture and dado rails, in the same way as skirting boards, taking care not to damage the existing decorations.

Repairing a door that is binding

One of the most common problems that occur with doors is that they **bind**. A door can bind at several different points, as you can see in the illustration on page 269.

These binding problems are simple to fix.

- If the door is binding at the hinges, it is usually caused by either a screw head sticking out too far or by a screw that has been put in squint. In these cases, screw the screw in fully or replace it straight. If the hinge is bent, fit a new one.
- If the door is binding on the hanging stile, it may not have been back-bevelled when hung. In this case, take the door off, remove the hinges, plane the door with a back bevel and re-hang it. Alternatively, it may be expansion or swelling, due to changes in temperature, that is causing the problem. In this case, take off the door, plane it and re-hang it.

> **Remember**
>
> When replacing mouldings, you may need to rub down the existing moulding work and re-apply a finish to both the old and the new to get a good match

> **Safety tip**
>
> Take care if gripper rods are fitted in front of the skirting, as this will make replacing the skirting more difficult and more dangerous

> **Key term**
>
> **bind** – a door 'binds' when it will not close properly and the door springs open slightly when it is pushed closed

- If the door is binding at the stop on the hanging side, the fitter may not have left a 1–2 mm gap between the stop and the door to allow for paint, in which case the stop will have to be moved. If there is a door frame rather than a door lining, there is no stop to remove. Simply plane the rebate using a rebate plane.

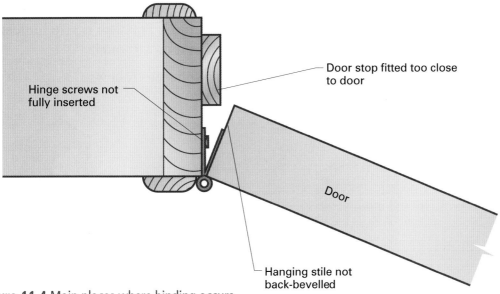

Hinge screws not fully inserted

Door stop fitted too close to door

Door

Hanging stile not back-bevelled

Figure 11.4 Main places where binding occurs

Repairing damaged door frames and windowsills

Provided they are painted or treated occasionally, exterior door frames and windows should last a long time, but certain areas of frames and windows are more susceptible to damage than others. The base of a door frame is where water can be absorbed into the frame. With windows, the sill is the most likely to suffer damage as water can sit on the sill and slowly penetrate it.

With damaged areas like these, the first thing to consider is whether to repair the damage or replace the component. With door frames replacement may be the best option, but you must take into account the extra work required to make good. Repairing the frame may cost less and require less work. With a rotten windowsill, replacement involves taking out the entire window so that the new sill can be fitted. Repair may be the better option.

The choice usually comes down to a balance of cost and longevity. As the experienced tradesperson, you must help the client choose the best course of action.

In this section we will look at repairing the base of a door frame and a windowsill.

Repairing a door frame

The base of the door frame is most susceptible to rot as the end grain of the timber acts like a sponge, drawing water up into the timber (this is why you should always treat cut ends before fixing).

First check that the timber is rotten. Push a blunt instrument such as a screwdriver, into the timber. If it pushes into the timber, the rot is evident; if it does not, the frame is fine. Rot is also indicated by the paint, or finish, flaking off or a musty smell.

If there is rot, repair it by using a splice, as follows.

Decay or defect
(a) Make cuts in the frame

(b) Remove the defect

(c) Scarfed splice

(d) Splice complete

Figure 11.5 Splicing a door frame

Step 1 Make 45 degree cuts in the frame with the cuts sloping down outwards, to stop surface water running into the joints.

Step 2 Remove the defective piece and use a sharp chisel to chop away the waste, forming the scarf.

Step 3 Use either a piece of similar stock or, if none is available, a square piece planed to the same size and shape as the original, making sure the ends are thoroughly treated before fitting to prevent a recurrence.

Step 4 Fix the splice in place, dress the joint using sharp planes and make good any plaster or render.

This method can also be used to repair door frames damaged by having large furniture moved through them, but the higher up the damage on the frame, the more likely the frame will need to be replaced rather than repaired.

Repairing a windowsill

As with a door frame, this method uses a splice, as follows:

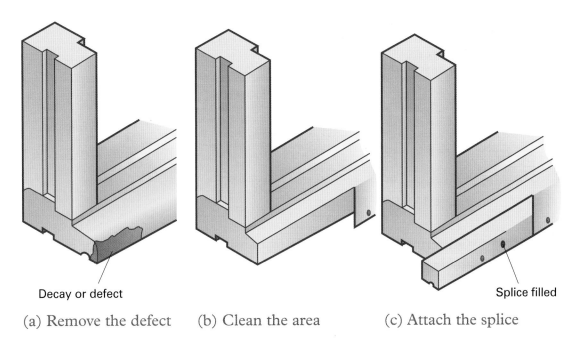

Decay or defect

Splice filled

(a) Remove the defect   (b) Clean the area   (c) Attach the splice

Figure 11.6 Splicing a windowsill

Step 1 Make a cut at 45 degrees, then carefully saw along the front of the windowsill to remove the rotten area.

Step 2 Clean the removed area using a sharp chisel and then, as before, use either a piece of similar stock or, if none is available, a square piece planed to the same size and shape as the removed area.

Step 3 Give all cut areas a thorough treating with preservative, then attach the splice to match the existing windowsill shape and paint or finish to match the rest.

K2. Know how to replace gutters and downpipes

Guttering is a vital part of a roof and must be maintained, and replaced if necessary, to ensure that the property remains waterproof. The purpose of guttering, which is attached to the fascia board, is to channel rainwater efficiently from the roof towards a downpipe, which in turn carries it down to ground level and then into the drainage system. This system helps to stop the walls of the house becoming soaked, causing problems with dampness.

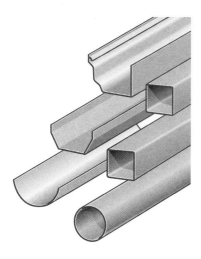

Figure 11.7 Types of gutter

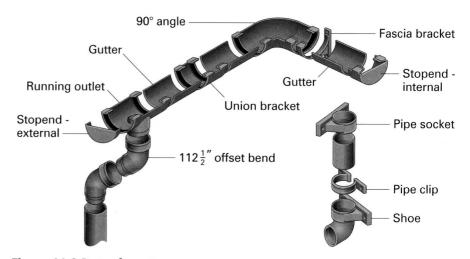

Figure 11.8 Parts of a gutter

90° angle

Gutter

Running outlet

Stopend - external

112½″ offset bend

Union bracket

Gutter

Fascia bracket

Stopend - internal

Pipe socket

Pipe clip

Shoe

Figure 11.9 Gutters must be properly maintained to prevent water damage to a building

Guttering used to be made from asbestos or cast iron, but is now made from pressed steel or, most commonly, uPVC.

The most common faults found in guttering are leaks. These are usually caused when the guttering is clogged up with leaves and dirt washed down the roof by rain. Regular clearing of the guttering will solve most problems.

In some cases, clearing the gutter may not be enough and occasionally the guttering may need to be replaced. Possible causes are rusted steel or cracked pipes and, with uPVC guttering, damaged seals or brackets may mean that they have to be replaced.

Doing the work

When doing any work at height, you must ensure that a suitable scaffold is used – ladders are not ideal for replacing guttering or doing longer jobs.

The first thing to do is to assess the damage, as there is no need to replace the entire system if only one joint is leaking. Asbestos guttering must be removed by a specialist firm in line with COSHH regulations. With cast-iron guttering, which is usually connected via bolts, finding replacements may prove difficult, so in some instances it may be better to replace the entire system with uPVC.

As uPVC is the most common type, we will look at the replacement of this type of guttering.

Note that various guttering profiles and colours are commercially available – and some older systems may even be obsolete. It is therefore important to check that the parts are available to repair the system you have, otherwise it will all have to be replaced.

Once the damaged area has been identified, the next thing is to remove the old guttering. uPVC guttering is fixed by screwing clips to the fascia, which the guttering clips into. Unclipping the guttering allows it to be removed safely and the replacement gutter can be clipped straight back into place.

If the whole gutter is to be replaced, first unclip all the guttering. Then either fix new brackets and clip in the new guttering, or clip the new guttering into the old clips.

Where one length of guttering meets another, a corner, a downpipe or a stop end, there are special brackets the gutter clips into.

Figure 11.10 Gutters clogged with leaves and dirt can cause leaks

The stop ends and corners simply clip onto the guttering, but the downpipe connector and straight connectors should be screwed to the fascia. If the whole system is to be replaced, you should screw the brackets and connectors to the fascia first, remembering to ensure that they are in line and running slightly downhill to where the water outlet is. Once the brackets are fitted, you can cut the guttering if needed and clip it into place, ideally running a bead of silicon along any joints.

The downpipe should be attached to the wall via brackets.

Figure 11.11 Gutters are available in a variety of profiles and colours

K3. Know how to replace window sash cords

As you saw in Unit 2008, box sash windows use weights attached to cords that run over a pulley system, to hold the sashes open and closed. The sash cord will eventually break through wear and tear and will need to be replaced. This is not a large job so, if one cord needs attention, it is most cost-effective to replace them all at the same time.

There are various ways to replace sash cords. The following method is quick and easy and is done from the inside.

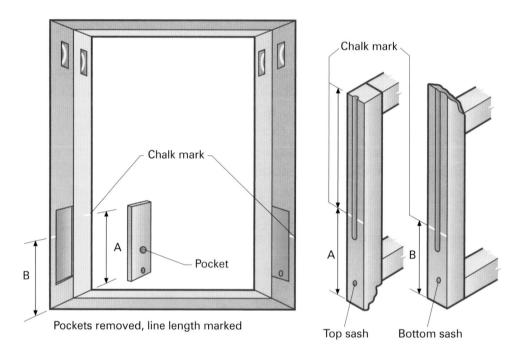

Figure 11.12 Marking sash lengths

Step 1 Remove the staff bead, taking care not to damage the rest of the window. Cut the cords supporting the bottom sash, lowering the weight gently to avoid damaging the case. Take out the bottom sash, removing the old nails and bits of cord and put to one side.

Step 2 Pull the top sash down and cut the cords carefully. Remove the parting bead and then the top sash, again removing the old nails and bits of cord, and put to one side.

Step 3 Remove the pockets and take out the weights, removing the cord from them and laying them at the appropriate side of the window.

Step 4 Slide the top sash into place at the bottom of the frame and, using chalk, mark the position of the sash cord onto the face of the pulley stiles (see distance A in Figure 11.12). Mark the bottom sash in the same way (distance B). Put the sashes to one side.

Step 5 Attach a **mouse** to the cord end, making sure it is long enough for the weighted end to reach the pocket before the sash cord reaches the pulley.

Step 6 Cord the window. If only one cord is to be replaced, you can do this by feeding one pulley, then cutting the sash to length, re-attaching the end of the mouse to the sash cord and then feeding the next pulley. If you are replacing more than one cord, it is more efficient to feed the pulleys in succession and cut the cords afterwards.

In Figure 11.14 you can see one method for cording the window.

Feed the cord through the top left nearside pulley (1), out through the left-hand pocket, in through the top left far side pulley (2), then back through the same pocket. Then feed it through the top right nearside pulley (3), out through the right-hand pocket, in through the top right far side pulley (4) and back out of the right-hand pocket. Remove the mouse and attach the right-hand rear sash weight to the cord end (5).

Key term

Mouse – a piece of strong, thin wire or rope with a weight, such as a piece of lead, attached to one end and used to feed the sash cord over the pulleys in a box sash window

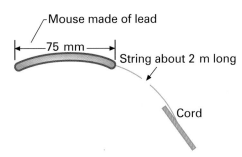

Figure 11.13 Mouse

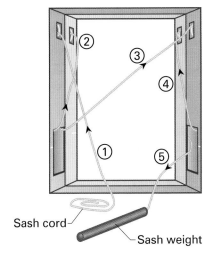

Figure 11.14 Cording a window

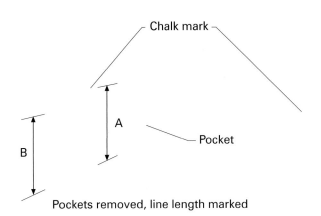

Step 7 Working on the right-hand rear pulley, pull one weight through the pocket into the box and up until it is just short of the pulley. Lightly force a wedge into the pulley to prevent the weight from falling.

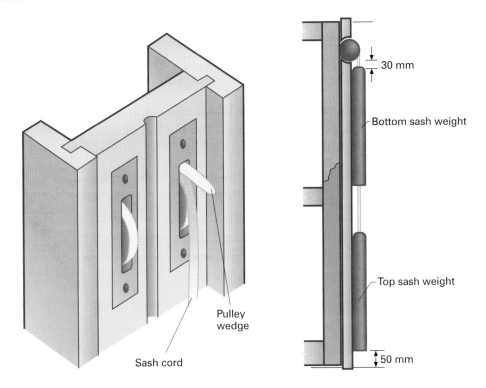

Figure 11.15 Wedged pulley

Figure 11.16 Sash weights at 50 mm and 30 mm

Pull the cord down the pulley stile and cut it to length, 50 mm above the chalk line. When the window is closed and the weight falls back down to the bottom of the box, the weight will stop 50 mm from the bottom of the frame as shown in Figure 11.15.

Step 8 Tie the other right-hand weight to the loose end of the cord. Still on the right-hand near-side, pull the cord again so that the weight is just short of the pulley and wedge it in place. Cut the cord to length, this time 30 mm longer than the chalk line. When the window is closed the bottom sash weight will be 30 mm away from the pulley, as shown in Figure 11.16.

Step 9 Fix the right-hand pocket back in place, then repeat Steps 7 and 8 for the left-hand side. Now all the pulleys are wedged and all the cords cut to length.

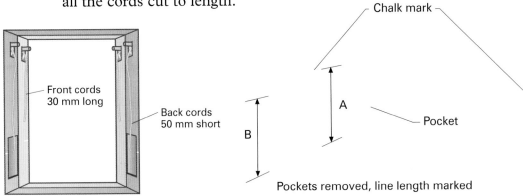

Figure 11.17 All pulleys wedged and cords cut to length

Pockets removed, line length marked

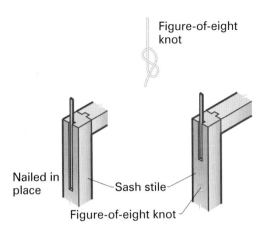
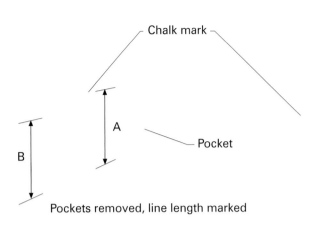

Figure-of-eight knot

Nailed in place

Sash stile

Figure-of-eight knot

Chalk mark

A

B

Pocket

Pockets removed, line length marked

Figure 11.18 Two methods for fixing cord to sashes

Step 10 Fit the sashes, starting with the top sash. Fix the cord to the sash by either using a knot with a tack driven through it, or a series of tacks driven through the cord. Take care not to hamper the opening of the window and do not use long nails as they will drive through the sash stile and damage the glass.

Step 11 Once the cord is fitted to both sides of the sash, slide the sash into place and carefully remove the wedges. Now test the top sash for movement and then re-fit the parting bead to keep the top sash in place. Fit the bottom sash as in Step 10, test it and re-fit the staff beads to secure the bottom sash in place.

Test the whole window by sliding both sashes up and down. Finally, touch up any minor damage to the staff and parting beads.

K4. Know how to make good plaster, paintwork and brickwork

During repair work there is always a chance that the interior plaster or brickwork may get damaged. Rather than call in a specialist most tradespeople will repair the damage themselves.

There are various types of plaster available:

- **browning plaster** – a backing coat plaster; this is normally grey or pink in colour. Browning is suitable for use on surfaces such as common bricks and building blocks, or other surfaces which are absorbent. This should be applied in layers of about 10 mm in thickness

Remember

Attempt only minor repairs unless you are fully trained. Otherwise you could end up doing more harm than good!

Functional skills

Working with mixes of plaster and paint will require you to practice FM 1.1.1 identify and select mathematical procedures. You will also need to practice FE 1.2.3 reading different texts and taking appropriate action, e.g. respond to advice/instructions.

Browning sets in about 1½- 2 hours.

- **Bonding plaster** – is more versatile than browning. Bonding can be used on dense, not very absorbent materials such as engineering bricks, or surfaces that have been treated with PVA. This should be applied in layers of about 8 mm in thickness

  Bonding sets in about 1½- 2 hours.

- **Finishing plaster** – is applied over the top of either bonding/ browning or over plasterboard. The multi-finish from British Gypsum is probably the best as it is suitable for most surfaces. Board finish is also available which is only for use on plasterboard and cannot be used on browning or bonding plaster

  Finishing plaster is normally applied as a final coat and is about 2 mm thick.

  Drying time will depend on the room temperature. Finishing plaster needs polishing with a float when it starts to go off, so don't leave it too long before polishing!

- **Undercoat plaster** – high impact resistance and quicker drying surface. Suitable for application by hand or mechanical plastering machine to most masonry backgrounds.

Figure 11.19
Browning plaster

Figure 11.20
Bonding plaster

Figure 11.21
Finishing plaster

Figure 11.22
Undercoat plaster

Repairing plasterwork

Plaster damage most often occurs when windows or door frames are removed – the plaster cracks or comes loose – and is simple to fix with either ready-mixed plaster (ideal for such tasks) or traditional bagged plaster, which is cheaper.

Bagged plaster needs to be mixed with water and stirred until it is the right consistency. Some people prefer a thinner mix as it can be worked for longer, while others prefer a consistency more like thick custard as it can be easier to use.

Whichever type of plaster you use, the method of application is the same.

First, clean the area to remove any debris or loose materials. Then brush the affected area with a PVA mix, watered down so it can be applied by brush. This acts as a bonding agent to help the plaster adhere to the wall.

Next, use a trowel or float to force the plaster into the damaged area. If it is quite deep you may have to part-fill the area and leave it to set, then put a finish skim over it. Filling a deep cavity in one go may result in the plaster running, leaving a bad finish.

Figure 11.23 Plaster being applied and trowelled

When the plaster is almost dry, dampen the surface with water and use a wet trowel to skim over the plastered area, leaving a smooth finish. Once the plaster is dry, the area can be re-decorated.

Repairs to exterior render are done in the same way, using cement instead of plaster.

Repairing brickwork

With repairs to brickwork, one of the first problems is finding bricks that match the existing ones, especially in older buildings. After that, the method is as follows.

Mortar is used in bricklaying for bedding and jointing the bricks when building a wall. Mortar is made of sand, cement, water and plasticiser. The mortar must be 'workable' to allow the mortar to roll and spread easily. The mortar should hold onto the trowel without sticking.

Step 1 Make a hole in the centre of the area to allow the removal of the bricks. Bricks rarely come out whole, but care must be taken not to damage the surrounding bricks.

Step 2 Remove the bricks and carefully clean away the old mortar using a cold chisel.

Step 3 Lay a mortar bed on the base of the opening.

Step 4 Place the first two bricks, making sure mortar is applied to all the joints.

Step 5 Place the last brick, again making sure there is mortar between all the joints, then use a pointing trowel to point in the new bricks to match the others.

1

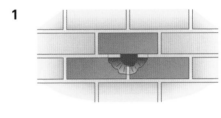

2

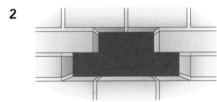

3

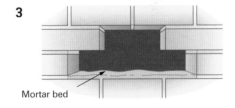

Mortar bed

4

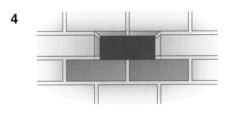

5

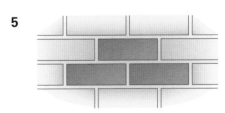

Figure 11.24 Stages in repairing brickwork

Mixing mortar by hand

If mixing by hand, the materials should be gauged first into a pile with the cement added. The cement and sand should then be 'turned' to mix the materials together.

The pile should be turned a minimum of three times to ensure the materials are mixed properly. The centre of the pile should be 'opened out' to create a centre hole. Gradually add the water, mixing it into the sand and cement, making sure not to 'flood' the mix. Turn the mix another three times, adding water gradually to gain the required consistency.

Mixing mortar by machine

Mixing by machine can be carried out by using either an electric, petrol or a diesel mixer. Always set the mixer up on level ground. If using an electric mixer the voltage should be 110 V and all cables and connections should be checked before use for splits or a loose connection. Cables should not be in contact with water and the operation should not be carried out if it is raining.

If using a petrol or diesel mixer, make sure the fuel and oil levels are checked and topped up before starting. If using the mixer for long periods, the levels should be checked regularly to ensure they do not run out. Gauge the materials to be used, and fill the mixer with approximately half of the water required (add plasticiser if being used). Add half the amount of cement to the water and add half of the sand. Allow to mix, and then add the remaining cement, then sand. Add more water if required, allowing at least two minutes for the mix to become workable and to ensure all the materials are thoroughly mixed together.

Once the mix has been taken out of the mixer, part fill the mixer with water and allow the water to run for a couple of minutes to remove any mortar sticking to the sides. If the mixer will not be used again that day, it should be cleaned thoroughly, either using water (and adding some broken bricks to help remove any mortar stuck to the sides) or ballast and gravel (which should then be cleaned out and the mixer washed with clean water). This will keep the mixer drum clean and any future materials used will not stick to the drum sides so easily.

Repairing paintwork

There are several reasons you might need to repair paintwork. They range from chipped or scratched paintwork to mould or damp.

The type of paintwork that you are repairing will depend on the general process, but with any paintwork it is important to ensure the preparation is done right.

Making paintwork repairs to walls and ceilings

Walls and ceilings are usually covered with water-based emulsion paint. If the walls are damaged through damp or mould it is important to first find and fix the cause or any repairs will need repeating until the underlying problem has been dealt with. Any scratches to walls or ceilings will need to be filled with a suitable filler and sanded flush.

Once the damaged or blemished area has been repaired it is best to give the walls a very light sand to provide a base for the next coats. Trying to just paint the repaired area will result in the new paint not matching exactly, so you should paint the whole wall or – if needed – the whole room. You will need at least two coats to ensure a decent finish.

Making paintwork repairs to mouldings/timber components

Timber components are usually coated with an oil-based paint. When repairing chips and scratches in existing paintwork, the repair process is similar to that used on walls and ceilings – the area is sanded and filled as needed.

Once the repair work has been completed, the whole area will need to be painted. Sometimes an undercoat is required. If so, this will be finished off with a top coat or gloss.

New timber will need to be primed if it has been used to replace damaged timber or as a scraf repair. Once new timber has been primed, it will need an undercoat and a top coat. As with paintwork repairs to walls or ceilings, you will need to re-paint the whole component – and possibly adjoining components – to ensure a matching finish.

FAQ

Can I do a scarf repair on skirting or architrave?

Yes, but it is often easier to replace mouldings than to repair them.

Should I go on a plastering course to help with repairs?

This is not necessary as the repairs you would be doing are not large, but if you feel that it would benefit you, then yes.

Check it out

1   Give a reason why mouldings may need to be replaced.
2   What is the name of the term used for cutting in a repair piece at 45 degrees?
3   Name the component attached to the end of a sash cord which is used to help feed the cord through the pulleys.
4   Why should trussed rafters not be repaired by a non-specialist?
5   What is the main cause of rot in rafters?
6   What is the first step to take before applying plaster?
7   Why is it important to check the guttering profile and colour first?
8   When are special clips for guttering needed?
9   Describe the following types of plaster:
    a browning
    b hardwall
10 Name four types of insects that attack timber.
11 Name three non-pressure methods of applying preservatives.

Getting ready for assessment

The information contained in this unit, as well as continued practical assignments that you will carry out in your college or training centre, will help you with preparing for both your end of unit test and the diploma multiple-choice test. It will also aid you in preparing for the work that is required for the synoptic practical assignments.

The information contained within this unit will aid you in learning how to identify and calculate the materials and equipment required to carry out maintenance.

You will need to be familiar with:

- preparing the materials for carrying out a range of maintenance tasks
- repairing mouldings
- repairing doors and windows
- replacing gutterings and down pipes
- replacing sash cords
- making good plaster, paintwork and brickwork.

This unit will have made you familiar with the methods used to maintain a number of the parts of a building. For example, for learning outcome three you have learned about the different types of gutters and downpipes, the materials that can be used to construct these and how they can become damaged. You will need to use this information on site to quickly see and identify problems with guttering and decide the best method to use in replacing them.

Before you start work on the synoptic practical test it is important that you have had sufficient practice and that you feel that you are capable of passing. It is best to have a plan of action and a work method that will help you. You will also need a copy of the required standards, any associated drawings and sufficient tools and materials. It is also wise to check your work at regular intervals. This will help you to be sure that you are working correctly and help you to avoid problems developing as you work.

Your speed at carrying out these tasks will also help you to prepare for the time limit that the synoptic practical task has. But remember, don't try to rush the job as speed will come with practice and it is important that you get the quality of workmanship right.

Always make sure that you are working safely throughout the test. Make sure you are working to all the safety requirements given throughout the test and wear all appropriate personal protective equipment. When using tools, make sure you are using them correctly and safely.

Good luck!

CHECK YOUR KNOWLEDGE

1 What can make it difficult to replace skirting?
  a gripper rods being fitted.
  b radiator pipes in the way.
  c other pieces of skirting being fixed to it.
  d all of the above.

2 When a door doesn't close properly or springs back open when pushed closed, this is known as:
  a bonding.
  b binding.
  c bunding.
  d bending.

3 How much gap should be left between a door and the doorstop to allow for paint?
  a 0–1 mm.
  b 1–2 mm.
  c 2–3 mm.
  d 3–4 mm.

4 How many steps are involved when replacing sash cords?
  a 8
  b 9
  c 10
  d 11

5 What are the weights on a sash window usually made from?
  a plastic.
  b timber.
  c cast iron.
  d silver.

6 When repairing structural timber, how far into sound wood should you cut?
  a 500 mm.
  b 600 mm.
  c 700 mm.
  d 1000 mm.

7 Prior to applying plaster, walls must be coated with:
  a water.
  b PVA.
  c sand.
  d powder.

8 What can cause a door to bind?
  a screw heads in the hinges sticking out.
  b the door hasn't been back-bevelled on the hanging stile.
  c the door stop is too close to the door.
  d all of the above.

9 Dry rot can be identified by:
  a an unpleasant, musty smell.
  b the appearance of fruity bodies.
  c dry, crumbly timber.
  d all of the above.

10 When eradicating wet rot, the most important thing to do is:
  a treat the timber with fungicide.
  b remove the rotted timber.
  c remove the source of wetness.
  d block up all ventilators.

11 What is the name of the process in which preservative is forced into the timber under pressure?
  a spraying.
  b steeping.
  c empty cell.
  d dipping.

How to set up and operate a circular saw

The circular saw is a vital piece of equipment for any carpenter or joiner. However, using a circular saw can be very dangerous if you have not been adequately trained. This unit covers how to set up a circular saw, how to operate it, and how to identify common hazards associated with circular saws.

This unit also supports NVQ Unit VR13 Set Up and Use Circular Saws.

This unit also contains material that supports the five generic units in the TAP and use of circular saws is delivered and assessed throughout the TAP.

This unit will cover the following learning outcomes:

- How to set up fixed and transportable circular saws

- How to change saw blades

- How to cut timber and sheet material.

K1. Know how to set up fixed and transportable circular saws

Knowing how to set up a circular saw involves more than just getting a piece of equipment ready to use. Due to the potential dangers of working with a circular saw, an operative capable of setting up the machine must also know about safety legislation, common hazards and how to proceed if any are identified, safety aids for use on the saw and the importance of dust extraction.

Legislation

As with all woodworking machines, circular saws can be dangerous unless handled properly. Paying close attention and taking care while you work will help you to avoid accidents. The following is a brief explanation of the main regulations governing the use of woodworking machines.

Compared to other industries, woodworking accounts for a large proportion of accidents. Woodworking machines often have high-speed cutters and many cannot be fully enclosed owing to the nature of the work they do.

The use of woodworking machines was originally governed by the Woodworking Machine Regulations 1974. The introduction of the Provision and Use of Work Equipment Regulations (PUWER) 1992 superseded the 1974 regulations, although regulations 13, 20 and 39 were still in use until the PUWER regulations were updated in 1998.

The PUWER regulations are explained in more detail in Unit 2002, but below are a few of the items relating to the safe use of all woodworking machines. Safety regulations relating specifically to a particular machine are noted in dedicated sections.

At the end of this section you should be aware of your responsibilities in operating circular saws and the main regulations that govern their use.

There have been four main pieces of legislation which govern the use of circular saws:

- Woodworking Machines Regulations 1974
- Provision and Use of Work Equipment Regulations 1992
- Provision and Use of Work Equipment Regulations 1998
- Safe Use of Woodworking Machinery Approved Code of Practice and Guidance.

When the PUWER were introduced in 1992, most of the 1974 regulations were revoked. The PUWER were then reviewed in 1998, at which time the rest of the 1974 regulations were replaced by the Safe Use of Woodworking Machinery Approved Code of Practice. This code of practice gives employers further information on how to comply with the PUWER for woodworking machinery.

General safety requirements

Safety appliances

Safety appliances such as push sticks / blocks and jigs must be designed so as to keep the operator's hands safe. More modern machines use power feed systems, eliminating the need for an operator to go near the cutting action. Power feed systems should be used wherever possible. In the absence of a power feed system, the appropriate push sticks/blocks must be used.

Working area

An unobstructed area is vital for the safe use of woodworking machines. The positioning of any machine must be carefully thought through to allow the machine to be used as intended. In a workshop environment, where there are several machines, the layout should be arranged so that the materials follow a logical path. Adequate access routes between machines must be kept clear and there should also be a suitable storage area next to each machine to store materials safely without impeding the operator or others.

Floors

The floors around machines must be flat and kept in a good condition. They must also be kept free from debris such as chippings, waste wood and sawdust. Any electricity supply, dust collection ductwork, etc. must run above head height or be set into the floor in such a way that it does not create a trip hazard. Polished surfaces must be avoided and any spills must be mopped up immediately. Non-slip matting around a machine is preferred, but the edges must not present a trip hazard.

Lighting

All areas must be adequately lit, whether by natural or artificial lighting, to ensure that all machine set-up gauges and dials are visible. Lighting must be strong enough to ensure a good view of the machine and its operations and lights must be positioned to avoid glare and without shining into the operator's eyes.

Heating

The temperature in a workshop should be neither too warm nor too cold and the area should be heated if needed. A temperature of 16°C is suitable for a workshop.

Controls

All machines should be fitted with a means of isolation from the electrical supply separate from the on / off buttons. The isolator should be positioned so that an operator can access it easily in an emergency. Ideally there should be a second cut-off switch accessible by others in case the operator is unable to reach the isolator. Machines must be fitted with an efficient starting / stopping mechanism, in easy reach of the operator. Machines must always be switched off when not in use and never left unattended until the cutter has come to a complete standstill.

Braking

All new machines must be fitted with an automatic braking system that ensures that the cutting tool stops within 10 seconds of the machine being switched off. Older machines are not required to have this, but PUWER states that all machines must be provided with controls that bring the machine to a controlled stop in a safe manner. The approved code of practice calls for employers to carry out a risk assessment to determine whether any machine requires a braking system to be fitted and includes a list of machines for which braking will almost certainly be needed. If you are unsure, contact your local Health and Safety Executive (HSE).

Training

No person should use any woodworking machinery unless they have been suitably trained and deemed competent in the use of that machine.

Maintenance

The machines should be maintained as per the manufacturer's instructions and should be checked prior to use every day, with the inspection and any findings recorded in a log.

Regulations specific to circular saws

The code of practice gives the following guidance specific to the use of circular saws:

- While operating the saw, the operator's hands should never be in line with the saw blade.

- A riving knife should be used to reduce the risk of kickback of the work piece. It should be fixed securely below the table and positioned directly behind and in line with the blade.

- The riving knife should be kept adjusted so that it is within 8 mm of the blade at table level.

- For saw blades 600 mm in diameter, or less, the vertical distance between the top of the blade and the top of the riving knife should not exceed 25 mm.

- For saw blades more than 600 mm in diameter, the riving knife should be at least 225 mm above the top of the blade.

- When removing the cut piece from between the saw and the fence, you should always use a push stick (unless the cut piece is more than 150 mm wide).

- When using a push stick, move your left hand to a safe position along the plate of the saw, to avoid an accident if the piece moves unexpectedly.

- Every circular saw should have marked, or displayed, on it the diameter of the smallest saw blade that should be used.

- A circular saw should not be used for ripping, unless the saw blade projects through the upper surface of the work piece at all times.

- If you use a circular saw for grooving, you should ensure that special guards are used to prevent access to the saw blade above the table.

Procedure for dealing with faults

If you discover a fault with the machine then the first step is to isolate the equipment and inform all other users that the machine is unsafe. Next, inform your supervisor so that trained people can fix the machine. Any hazards should be dealt with as soon as possible and hazards that you can not deal with must be reported so that the appropriate action can be taken.

Dust collection and extraction

Woodworking machines must be fitted with an efficient means of collecting the dust or chippings produced during the machining process. Dust extraction on circular saws is usually by way of a hose connected to a vacuum with bags that collect the dust. Dust removal is vital, not only because of the damage that can be caused by breathing it in, but also because fine dust can combust and is a fire risk.

Functions of the saw

The main function of a hand-fed circular saw is to **re-saw** timber. This involves ripping with the grain and can be divided into the following four operations:

- flatting
- deeping
- bevel cutting
- angled cutting.

Flatting

Flatting is the term used to describe the cutting of timber to the required width on a hand-fed circular saw. See Figure 12.1.

Deeping

Deeping is the term used to describe the cutting of timber to the required thickness on a hand-fed circular saw. See Figure 12.2.

Figure 12.1 Flatting

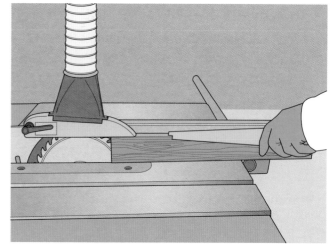

Figure 12.2 Deeping

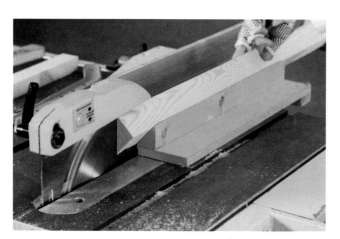

Figure 12.3 Jigs for bevel cutting

Bevel cutting

If a bevel is required to be cut on to a section of timber using a hand-fed circular saw, then the saw operator will make a jig that will safely carry out the operation. These jigs are known as 'bed pieces' and 'saddles'. See Figure 12.3.

Angled cutting

Angled cutting refers to cutting a piece of timber at an angle. The method is the same as cutting a bevel and, once again, the operator would make the appropriate jig for the task.

Components of a circular saw

To operate a hand fed circular saw you must be able to identify the main parts of the machine and understand how they function. This section looks at all the main parts in turn.

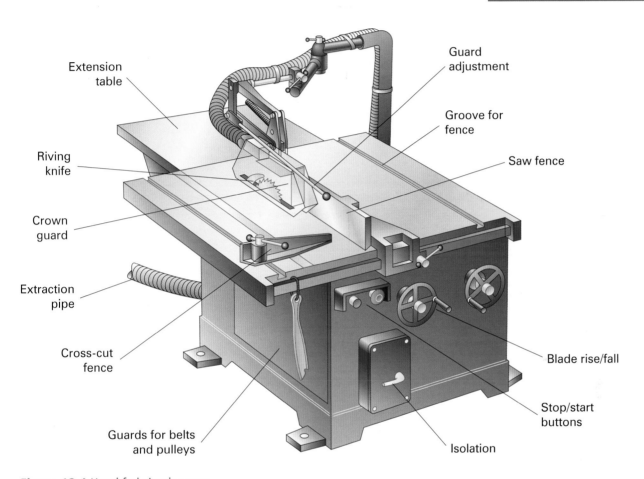

- Extension table
- Guard adjustment
- Groove for fence
- Riving knife
- Saw fence
- Crown guard
- Extraction pipe
- Cross-cut fence
- Blade rise/fall
- Guards for belts and pulleys
- Stop/start buttons
- Isolation

Figure 12.4 Hand fed circular saw

Riving knife

Riving knives are made from spring-tempered steel. The two main functions of the knife are:

- to prevent timber binding on the saw after cutting
- to act as a guard to the back of the saw teeth.

Figure 12.6 illustrates the main parts.

To avoid the timber binding on the saw blade, the riving knife must be thicker than the saw blade.

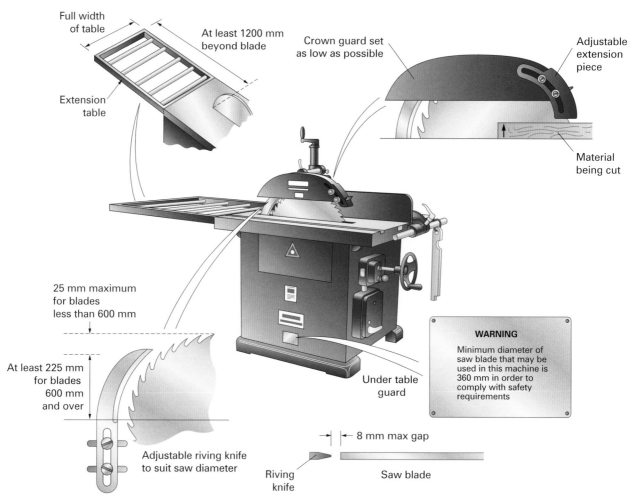

Full width of table

At least 1200 mm beyond blade

Extension table

Crown guard set as low as possible

Adjustable extension piece

Material being cut

25 mm maximum for blades less than 600 mm

At least 225 mm for blades 600 mm and over

Adjustable riving knife to suit saw diameter

Under table guard

WARNING

Minimum diameter of saw blade that may be used in this machine is 360 mm in order to comply with safety requirements

8 mm max gap

Riving knife

Saw blade

Figure 12.5 Riving knife parts

Figure 12.6 Circular saw showing crown guard, extension piece, mouthpiece, packings and finger plate

Crown guard and extension piece

To protect the operator from the top portion of the saw blade, a crown guard is fitted. This must cover the roots of the saw teeth. The crown guard is made of aluminium, steel or sometimes brass. On all hand-fed circular saws, the blade should be guarded above and below the bench, and flanges should provide protection on both sides.

The function of the extension piece is to guard the part of the saw between the crown guard and the timber. It is made of brass or steel. The extension piece must be as close as possible to the timber in order to prevent the operator's fingers from contacting the saw blade. The flange should be on the side away from the fence. Both flanges must extend beyond the roots of the saw teeth. The guards and extension pieces should be kept adjusted to maintain this.

Mouthpiece

The mouthpiece fits into the front portion of the packing box, between the packing and the front of the saw teeth. Its purpose is to prevent damage to the saw teeth and packing. It is made from a suitable hardwood or plywood. The mouthpiece is cut to an appropriate size to suit:

- the diameter of the saw being used
- the working height of the saw above the table.

Packings

The purpose of packings is to assist the saw to run true and they can be adjusted for tightness. They are made of gland felt or wood strips wound round with hemp.

Finger plate

When access to the saw blade is needed, the finger plate can be removed. The front section supports one side of the packing box. If the saw blade should wobble when running, damage to the teeth is prevented by a hardwood insert in the finger plate. Finger plates are made of cast iron.

Stop / Start buttons

The stop / start buttons on the machine must be easily accessible and work efficiently. Usually, the stop button will be a 'mushroom head' type with a twist release. To prevent the machine being started accidentally, the start button is normally recessed.

Extension table

An extension table is required to extend a minimum distance of 1200 mm behind the part of the saw blade that is running above the table and the full width of the table. This does not apply to movable machines with a maximum blade diameter of 450 mm.

The purpose of this requirement is to protect an operator 'pulling off' timber behind the saw bench. When pulling off, you should stand behind the table.

K2. Know how to change saw blades

It is very important to select the correct type of saw blade for the job. At the end of this section you should be able to:

- identify types of blades
- understand terms used for parts of a blade
- understand how a blade fits on to a machine.

Figure 12.7 Stop / start buttons

Figure 12.8 Extension table

Types of blade

Although a circular saw blade may seem quite simple in construction, each part must be set up correctly for efficient performance in use. There are two main types used:

- tungsten carbide tipped (TCT) blade
- sprung steel blade.

TCT blades are more durable than the more traditional sprung steel blade, as the latter need to have each individual tooth set properly to work effectively.

The teeth on rip saws must have chisel edges, sloping towards the wood. This is called positive hook (see below). More hook is needed for ripping softwood than hardwood.

The cutting edge of sprung steel blades quickly become dull when ripping abrasive timbers, so TCT blades are recommended when cutting these, as they have greater resistance to wear and blunting.

The use of a saw blade is determined by the type of grind and the spacing of the teeth. Crosscut saw blades generally have more teeth and are at a different angle to a rip saw blade, giving a neater finish. Hollow ground blades are ideal for laminates and delicate work, but they tend to be brittle. Combination plate blades are general purpose blades that can be used for most tasks.

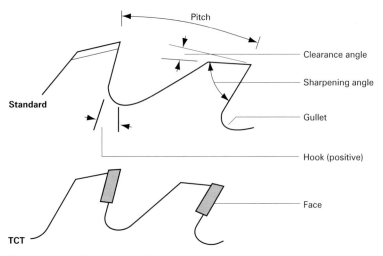

Figure 12.9 Circular saw blades

Saw blade terminology

Key terms used in relation to blades are:

- **back** – the back of the tooth
- **point** – the point of the tooth
- **heel** – the back edge of the top
- **pitch** – the distance between the point of two teeth
- **hook** – the angle of the front of the tooth
- **positive hook** – when teeth incline towards the timber, as required for ripping (20°–25°, maximum 30°, for softwoods; 10°–15°, maximum 20°, for hardwoods)

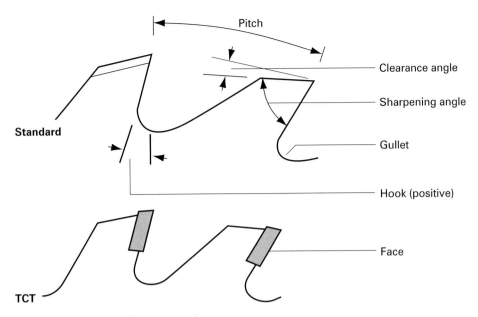

Figure 12.10 Parts of a saw tooth

- **clearance angle** – ensures the heel of the tooth clears the timber when cutting (15° is normal for softwood; 5°–10° for hardwoods)

- **top bevel** – angle across the top of the tooth (15° for softwoods; 5°–10° for hardwoods)

- **front bevel** – bevel across the face of the tooth (nil for a rip saw)

- **gullet** – space between each tooth to carry away sawdust

- **root** – the base of the tooth

- **kerf** – total width of the saw cut made by the blade (twice the set plus the thickness of the saw plate on a sprung steel blade; twice the overhang plus the thickness of the saw plate on a TCT blade)

- **set** – the amount on sprung steel blades that each tooth is bent, or 'sprung out', to give a clearance on the saw plate

- **overhang** – on TCT blades, the amount that the cutting edge is wider than the saw plate.

> **Did you know?**
>
> Many sawmills now use TCT blades because they stay sharper for longer, especially on abrasive timbers, and do not require set as the tips overhang the saw plate

Saw blade fitting

Saw blades are fitted on to a spindle between two flanges or collars. When a saw blade needs to be fitted or removed, the front collar can be taken off to allow access. The rear collar is fixed to the spindle. In some cases, circular saws will have a locating pin in the rear collar, which assists in the saw blade taking up the drive of the motor.

Care must be taken when replacing any saw blades and it is important to check the manufacturer's instructions. When

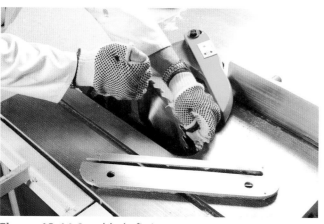

Figure 12.11 Saw blade fitting

changing the blade for one of a different size, or style, it is vital to check that the replacement blade is suitable. A saw should have a label on it stating the minimum / maximum size of blade to be fitted to the machine. This is important as the spindle on the saw runs at a certain speed to drive the blade. The blade runs at what is called the peripheral speed.

The life of a saw blade will vary greatly and depends upon the materials that it is used to cut. Some timbers which are very strong, such as oak, will dull a blade more quickly than timbers like Douglas fir. Man-made boards, such as chipboard or plywood, contain glues which will also blunt a blade.

K3. Know how to cut timber and sheet material

There are several preparations to be made, and factors to be considered, before using a hand-fed circular saw to rip timber. These include:

- preliminary checks
- correct setting of the fence
- selecting additional safety work pieces required to feed timber through the machine
- identifying the safest position for the operator to stand while feeding timber into the machine.

Preliminary checks

Prior to starting the machine the operator must ensure that:

- the machine is isolated from the power supply
- the saw blade is suitable for the job, that it is sharp and tensioned correctly
- the saw collars are clean and in good condition
- the saw blade is mounted correctly
- the riving knife is positioned correctly
- the finger plate is located correctly
- packings and mouthpiece are a good fit
- the height that the saw blade is set suits the depth of cut being made

- the saw blade runs free when slowly turned by hand
- the fence position is in line with the gullets of the saw at table level
- the extension piece and crown guard are correctly positioned
- the fine adjustment on the fence is working properly, that is not sticking, nor failing to operate on the thread
- dust extraction equipment is set up correctly
- additional safety work pieces are available
- the timber is free of nails, grit or other foreign bodies.

Setting the fence

The fence is a component part of a saw table, easily set to enable accurate cutting to the size of sheet materials or lengths of timber. The fence usually moves on a slide to the distance required between the blade and inside face of the fence. Some fences are fitted with a fine adjustment facility.

It is vital when setting the fence to ensure that the end of it does not exceed the position of the saw tooth gullets at table level. This is to eliminate the possibility of timber binding on the blade.

Determining the width of cut

The best way to set the width of the cut is to measure from the inside edge of the tooth to the fence. This will compensate for different amounts of set or overhang on the teeth of different saws. This is shown in Figure 12.12.

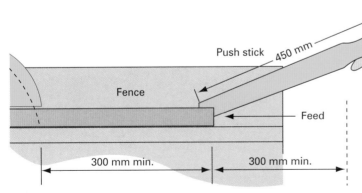

Figure 12.12 Method for setting the width of the cut

The rule attached to the table can be used for approximate setting, although the accuracy will vary according to the saw blade used. It is better to use a separate steel rule.

Figure 12.13 Using the rule to set the width of the cut

It is the operator who is ultimately responsible for proper adjustment of the guard on any woodworking machine

Operator position when operating the saw

Operators should always position themselves with one foot a comfortable distance in front of the other. This will ensure good balance and a strong stance. The operator should also make sure they are able to reach the emergency stop button easily.

If standing behind the machine to pull cut timber through, an operator should always make sure they are clear of the saw blade.

Figure 12.14 Best position for the operator

Figure 12.15 Operator pulling timber through

The operator should support the timber being sawn and move backwards as it is fed through the saw. The two saw operators should always work together and listen carefully to each other's instructions.

Final checks

Having carried out the preliminary checks given above, the saw can now be started and a trial cut made, after which the size of the cut can be checked and adjustments made, if necessary, to the machine settings.

If there is any wobble on the blade the cut will be wider than expected. The trial cut should be measured with a steel rule and adjustments made as necessary using the fine adjustment screw.

Did you know?

When flatting brittle timbers, and sheet materials such as plywood, 'spelching' can be minimised by having the minimum projection of saw teeth above the material. Spelching is the breaking out of fibres when wood, especially plywood, is cut

Illegal methods of use

It is dangerous to try to use a circular saw for operations for which it is not designed. In particular, the regulations state that they must not be used for cutting rebates, tenons, moulds or grooves. The use of a proper saw guard attachment is the only legal way of performing these tasks.

Working life

Nathan has been asked to cut some bevelled timber using a circular saw.

What checks should Nathan perform before operating the saw to make sure it is safe to use? Nathan will need to run through all the safety checks to ensure that the tool is not going to harm him or (potentially) other workers. How can Nathan produce bevelled timber with a circular saw? Do you think he needs any additional tools or equipment?

Can Nathan do this job on his own? Who else, if anyone, should Nathan think about involving or getting help from before he begins?

FAQ

What sign must be displayed on a circular saw?

The diameter of the smallest saw blade that can safely be used with the saw in question. A small diameter blade (less than 60 per cent of the diameter of the largest blade the saw can accommodate) will have a low peripheral blade speed and will cut inefficiently.

What type of saw blade would be most suitable for ripping abrasive timbers?

Tungsten carbide tipped (TCT).

What common tasks may be carried out on a circular saw bench?

Ripping, cross-cutting, bevelling and angle cuts.

While ripping timber, the blade produces blue smoke, and the operator finds it difficult to push the timber through. What might be the cause?

Possible errors include fence being incorrectly set, the saw blade being blunt or no set on the blade.

Check it out

1. Name two pieces of legislation that cover the use of circular saws.
2. State the four common cuts that can be made using a circular saw.
3. State the purpose of a riving knife.
4. State what distance should the extension piece be from the timber, and why.
5. What is the function of a mouthpiece?
6. Why is the start button normally recessed?
7. What is the minimum distance of an extension table from the back of a blade?
8. What is the recommended length of a push stick?
9. What does TCT stand for?
10. What is 'kerf'?
11. Describe 'set' in relation to saw blades.
12. Name five parts of a saw tooth.

Getting ready for assessment

The information contained in this unit, as well as continued practical assignments that you will carry out in your college or training centre, will help you with preparing for both your end of unit test and the diploma multiple-choice test. It will also aid you in preparing for the work that is required for the synoptic practical assignments.

The information contained within this unit will aid you in learning how to identify the components of a circular saw as well as the safe set up and operation.

You will need to be familiar with:

- how to set up fixed and transportable circular saws
- how to change saw blades
- how to cut timber and sheet material.

Using circular saws can be very dangerous if they are not used correctly. You will need to remember to follow health and safety guidance at all times to make sure you, and everyone around you, is safe while you work. For learning outcome one you have seen the potential hazards the exist when working with circular saws, as well the current legislation that governs the use of circular saws. You will need to be able to follow these guidelines to check circular saws for potential hazards such as faulty or missing guards, faulty or wrongly fitted tooling or damage to equipment. You will also need to be sure that you have set up the saw in line with the manufacturer's recommendations and current legislation.

Before you start work on the synoptic practical test it is important that you have had sufficient practice and that you feel that you are capable of passing. It is best to have a plan of action and a work method that will help you. You will also need a copy of the required standards, any associated drawings and sufficient tools and materials. It is also wise to check your work at regular intervals. This will help you to be sure that you are working correctly, and help you to avoid problems developing as you work.

Your speed at carrying out these tasks will also help you to prepare for the time limit that the synoptic practical task has. But remember, don't try to rush the job as speed will come with practice and it is important that you get the quality of workmanship right.

Always make sure that you are working safely throughout the test. Make sure you are working to all the safety requirements given throughout the test and wear all appropriate personal protective equipment. When using tools, make sure you are using them correctly and safely.

Good luck!

CHECK YOUR KNOWLEDGE

1 There should be sufficient lighting in a workshop to ensure that:
   a there is a good glare.
   b the dials and gauges are visible.
   c tinted safety goggles are needed.
   d all of the above.

2 No person should use woodworking machinery unless:
   a they are over 18.
   b they are over 21.
   c they have been trained.
   d their supervisor says so.

3 All circular saws should be fitted with a brake that stops the blade within:
   a 1 second.
   b 10 seconds.
   c 100 seconds.
   d 10 minutes.

4 The spreader that sits at the rear of the blade on a table saw and prevents binding is called the:
   a crown guard.
   b riving knife.
   c fingerplate.
   d fence.

5 What is the name for the distance between two teeth on a saw blade?
   a hook.
   b pitch.
   c gullet.
   d kerf.

6 The correct type of timber conversion is:
   a up and under.
   b over and out.
   c through and through.
   d round and round.

7 Boxed heart conversion is used when:
   a the best quality finish is required.
   b the timber needs to be converted cheaply.
   c the heart of the tree is rotten.
   d strength is important.

8 We season timber to:
   a reduce the moisture.
   b make the timber stronger.
   c allow the timber to be painted.
   d do all of the above.

9 The correct moisture content for external joinery is:
   a 18–20%.
   b 16–18%.
   c 14–16%.
   d 10–12%.

10 What type of defect occurs when timber is seasoned too rapidly and the outside dries, sealing the inside?
   a collapse
   b case hardening
   c cupping
   d cup shakes

Know how to produce setting out details for routine joinery products

As with all industries, technological advances within the construction industry have led to changes in the way work is done. While routine joinery products are now available ready-made and 'off-the-shelf', a good carpenter or joiner will have the skills to make doors, stairs and windows. Mass-produced items can save time and money, but not every job will suit factory sizes or delivery times. It is for these jobs that you will need to use the setting out skills taught in this unit.

This unit also supports NVQ Unit VR14W Produce Setting out Details for Routine Products.

This unit also contains material that supports TAP Unit 5 Produce Joinery Setting Out Details.

This unit will cover the following learning outcomes:

- How to interpret information for setting out
- How to select resources for setting out
- How to set out for bench joinery and site carpentry.

K1. Know how to interpret information for setting out

Understanding how to interpret setting out information is a vital tool in a carpenter or joiner's arsenal. A lack of comprehension here will affect your ability to work – if you are not able to take the setting out information and turn it into the expected item built to the right size and using appropriate materials, you will spend a lot of time re-doing work you have already done, wasting your time, the company's time, and, most importantly, the client's time.

Basic setting out

At the end of this section you will understand:

- the principles of a setting out rod and its uses
- the purpose of a cutting list.

Setting out rod

A setting out rod will usually be a thin piece of plywood, hardboard or medium density fibreborad (MDF), on which can be drawn the full size measurements of the item to be made. It is often painted white to improve the clarity of the drawing.

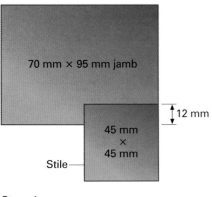

70 mm × 95 mm jamb
45 mm × 45 mm
Stile
12 mm
Stage 1

Figure 34.1 White setting out rod for small, four-pane sash

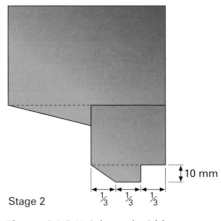

10 mm
⅓ ⅓ ⅓
Stage 2

Figure 34.2 Height and width sections

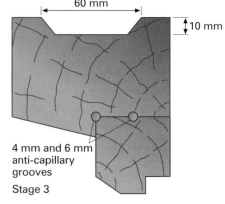

60 mm
10 mm
4 mm and 6 mm anti-capillary grooves
Stage 3

Figure 34.3 Rod with critical dimensions for a single panel glazed door

Rods can be used time and time again, simply by re-painting the surface upon completion of a task. If marked rods are to be kept for re-use they must be referenced and stored safely.

Upon receipt of scale drawings, specification and any on site measurements the **setter out** will produce a full size, horizontal and vertical section through the item by drawing it on a setting out rod. See Figure 34.1.

Elevations may also be drawn on setting out rods. This is particularly valuable for shaped or curved work, as the setter out can get a 'true' visual image of a completed joinery item.

Although rods are marked up full size, certain critical dimensions can be added as a check against any errors or damage to the rod. These are usually:

- **sight size** – the size of the innermost edges of the component (usually the height and width of any glazed components and, therefore, sometimes referred to as 'daylight size')
- **shoulder size** – the length of any member between shoulders of tenons
- **overall size** – the extreme length and width of an item.

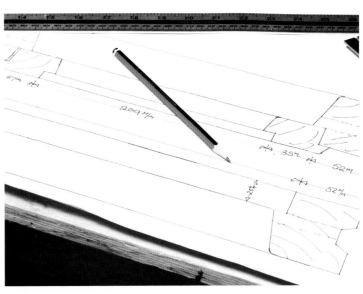

Figure 34.4 Rod marked up for a casement window

Developing drawn components

When producing workshop rods an inexperienced or apprentice joiner can sometimes have problems when building up a detailed section of timber. To overcome this, use the following step-by-step guidelines.

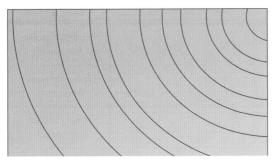

Figure 34.5 Step 1 Draw the components as a rectangular section

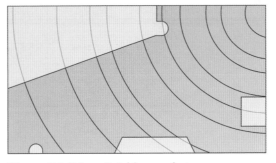

Figure 34.6 Step 2 Add any rebates, grooves and mouldings

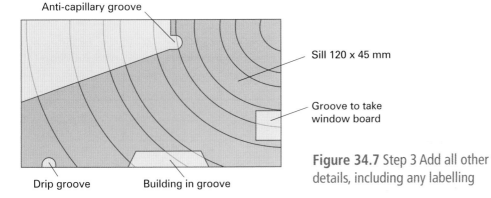

Anti-capillary groove

Sill 120 x 45 mm

Groove to take window board

Drip groove   Building in groove

Figure 34.7 Step 3 Add all other details, including any labelling

Cutting lists

Once the setting out rod has been completed the cutting list can be compiled. The cutting list is an accurate, itemised list of all the timber required to complete the job shown on the rod.

The cutting list will need to be referred to throughout the manufacturing process. It is, therefore, good practice to include the cutting list on the actual rod wherever possible.

Although there is no set layout for a cutting list, certain information should be clearly given in all lists. It should include:

- a reference for the setting out rod, i.e. rod number
- the date the list was compiled
- a brief job description
- the quantity of items required
- the component description (e.g. head, sill, stile, etc.)
- the component size, both sawn and finished (3 mm per face should be allowed for machining purposes)
- any general remarks.

An example cutting list is shown in Figure 34.8.

Timber cutting list						
Job description: Two panel door			Date: 8 Sept 2010			
Quantity	Description	Material	Length	Width	Thickness	Remarks
2	Stiles	S wood	1981	95	45	Mortise/groove for panel
1	Mid rail	"	760	195	45	Tenon/groove for panel
1	Btm rail	"	760	195	45	Tenon/groove for panel
1	Top rail	"	760	95	45	Tenon/groove for panel
1	Panel	Plywood	760	590	12	
1	Panel	"	600	590	12	

Figure 34.8 A cutting list

Check work for accuracy

Whatever brief you are given regarding work to be carried out, it is important that it is checked and confirmed to be accurate. Failure to check for accuracy will result in parts of, or whole, tasks being done incorrectly. This in turn can lead to the work having to be re-done, which can be very costly in both time and money.

Drawings should match each other, as well as the measurements of the area. It is of no use to anyone if the drawing shows a window to be made 2 metres high and the opening is only 1.9 metres high. It is also important to check that all drawings and specification meet the client's requirements. Sometimes drawings may have been changed slightly to meet regulations, and the client must be kept informed. Alternatively, a client might change their mind once they see a drawing, leading to slight alterations or variations.

If you discover any discrepancies it is vital that you inform your supervisor and any people who are associated with the task. If they continue to work to incorrect drawings the work they produce will be wrong and can prove to be costly.

K2. Know how to select resources for setting out

You may be familiar with the expression 'a good carpenter always has the right tools for the job'. What you might not know is that tools means more than just those of the hand and power varieties. Tools used in that context means equipment, and a carpenter and joiner needs to have the right timber, sheet material, or man-made board at his or her disposal. After all, the best equipped toolbox in the world is useless when there is nothing around to work with.

Identification of timber and materials

The identification of timber has been covered in Unit 2008, pages 132–135. As you become more skilled, you will find that you are working with a variety of materials, not all of them timber. Sheet materials are often used in joinery work.

The standard size for sheet material is 2400 × 1200 mm, although it is also available in 1800 × 1200 mm, 1800 × 900 mm, and 1200 × 600 mm, in increments of 300 mm. The thickness of the sheet varies in multiples of 3 mm, and so goes 3 mm thick, 6 mm thick, 9 mm thick, and so on.

Functional skills

Setting out will allow you to practice FM 1.2.1b interpret information from sources such as diagrams, tables, charts and graphs and FM 1.2.1c draw shapes. You will also cover FM 1.3.1 judge whether findings answer the original problem and FM 1.3.2 communicate solutions to answer practical problems.

Timber sizes vary greatly with all dimensions and can be made available or machined to any size. Lengths of timber usually start at 1200 mm and rise in increments of 300 mm. The width and thickness of timber are still usually referred to in imperial sizes such as 3" × 2" as opposed to 50 mm × 75 mm. These sizes are numerous by usually going up by 1" or 25 mm increments such as 2" × 1", 2" × 2", 3" × 2".

Whatever the work involved, the quality of the finished product depends on the skill of the operator in selecting and using the correct tools to cut, shape and assemble the right materials for the task.

K3. Know how to set out for bench joinery and site carpentry

There is a range of skills involved in knowing how to set out for bench joinery and site carpentry and in this section you will find an overview. Once you have selected the appropriate materials for the job, as covered in the previous section, you will need to choose the right tools for the task at hand, know which joints are suitable for the project and how to plan your work so that you complete the job on schedule.

Measuring and marking out tools

The main tools for measuring and marking out are:

- folding rules
- retractable steel tape measures
- metal steel rules
- pencils
- marking knife
- tri-square
- sliding bevel
- mitre square
- combination square
- gauges (marking, mortise and cutting gauges)
- callipers and dividers.

Folding rules

Folding rules are used in the joiner's shop or on site. They are normally one metre long when unfolded and made of wood or plastic. They can show both metric and imperial units.

> **Remember**
>
> Imperial measurements (e.g. yards, feet, inches) have been replaced by metric measurements (e.g. metres, millimetres), but most older items will have been constructed in imperial

Figure 34.9 Folding rule

Retractable steel tape measures

Retractable steel tape measures, often referred to as spring tapes, are available in a variety of lengths. They are useful for setting out large areas or marking long lengths of timber and other materials. They have a hook at right angles at the start of the tape to hold over the edge of the material. On better tapes this should slide, so that it is out of the way when not measuring from an edge.

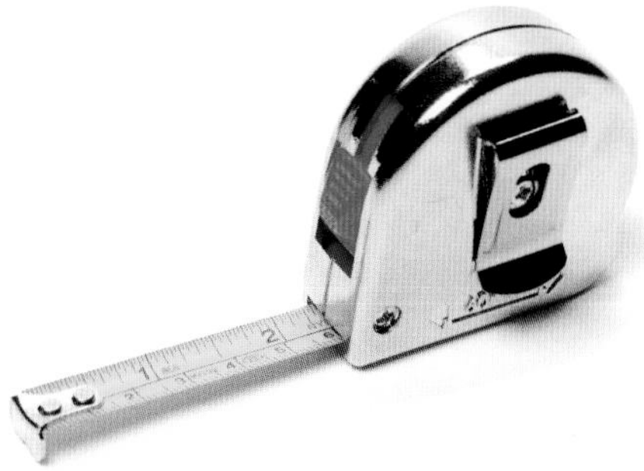

Figure 34.10 Retractable steel tape measure

> **Remember**
>
> Check the sliding hook regularly for signs of wear

Metal steel rules

Metal steel rules, often referred to as bar rules, are used for fine, accurate measurement work. They are generally 300 mm or 600 mm long and can also serve as a short straight edge for marking out. The rule can also be used on its edge for greater accuracy.

They may become discoloured over time. If so, give them a gentle rub with very fine emery paper and a light oil. If they become too rusty, replace them.

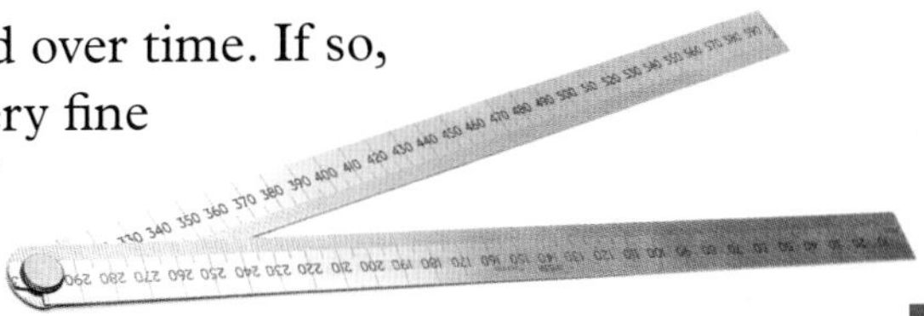

Figure 34.11 600 mm steel rule

Pencils

Pencils are an important part of a tool kit. They can be used for marking out exact measurements, both across and along the grain. They must be sharpened regularly, normally to a chisel-shaped point, which can be kept sharp by rubbing on fine emery paper. A chisel edge will draw more accurately along a marking out tool, like a steel rule, than a rounded point.

Pencils are graded by the softness or hardness of the lead. B grades are soft, H grades hard, with HB as the medium grade. Increasing hardness is indicated by a number in front of the H. Harder leads give a finer line, but are often more difficult to rub out. A good compromise for most carpentry work is 2H.

> **Did you know?**
>
> The 'lead' in pencils is actually graphite, a naturally occurring form of compressed carbon. If graphite is pressed much harder, it becomes diamond

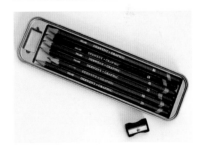

Figure 34.12 A variety of pencils

Marking knife

Marking knives are used for marking across the grain and can be much more accurate than a pencil. They also provide a slight indentation for saw teeth to key into.

Figure 34.13 Marking knife

Tri-square

Tri-squares are used to mark and test angles at 90° and check that surfaces are at right angles to each other.

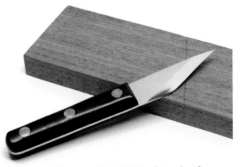

Tri-squares should be regularly checked for accuracy. To do this, place the square against any straight-edged spare timber and mark a line at right angles. Turn the square over and draw another line from the same point. If the tool is accurate the two lines should be on top of each other.

Figure 34.14 Tri-square

Figure 34.15 Sliding bevel

Sliding bevel

The sliding bevel is an adjustable tri-square, used for marking and testing angles other than 90°. When in use, the blade is set at the required angle and locked by either a thumbscrew or set screw in the stock.

Mitre square

The blade of a mitre square is set into the stock at an angle of 45° and is used for marking out a mitre cut.

Figure 34.16 Mitre square

> **Remember**
>
> Do not over-tighten thumbscrews on a sliding bevel as they may snap

> **Remember**
>
> All squares should be checked for accuracy on a regular basis

Combination square

A combination square does the job of a tri-square, mitre square and spirit level all in one. It is used for checking right angles, 45° angles and also that items are level.

Figure 34.17 Combination square

Gauges

Gauges are instruments used to check that an item meets standard measurements. They are also used to mark critical dimensions such as length and thickness.

Marking gauge

A marking gauge is used for marking lines parallel to the edge or end of the wood. The parts of a marking gauge include stem, stock, spur (or point) and thumbscrew. A marking gauge has only one spur or point.

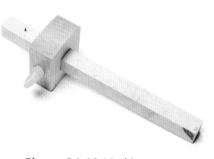

Figure 34.18 Marking gauge

Mortise gauge

A mortise gauge is used for marking the double lines required when setting out mortise and tenon joints, hence the name. It has one fixed and one adjustable spur or point. Figure 34.20 shows setting of the adjustable point to match the width of a chisel.

Figure 34.19 Mortise gauge

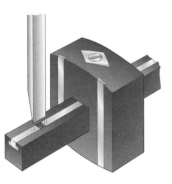

Figure 34.20 Setting mortise gauge to chisel blade width

Cutting gauge

The cutting gauge is very similar to the marking gauge, but has a blade in place of the spur. This is used to cut deep lines in the timber, particularly across the grain, to give a clean, precise cut (e.g. for marking the shoulders of tenons).

Callipers and dividers

Callipers and dividers enable accurate checking of widths and gaps. They can have a simple friction joint or knurled rod and thread. The latter are more accurate for repetitive work, as the width setting can be maintained.

Callipers are designed for either internal and external gaps. Although some come with a graduated scale, it is usually better to check measurements against a steel rule.

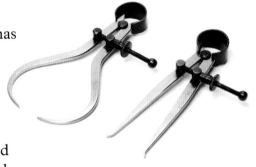

Figure 34.21 External and internal callipers

Organising work in sequence

One of the most important things to consider when organising work is the sequence of operations. If you get this wrong it can lead to work having to be re-done. When making windows, for example, you would not fully fit the glass and glazing bead before the window is installed as you may need to fix the window through the rebate where the glass sits. To do this once the window has been installed would mean having to remove the glass, fit the window and re-install the glass. This is extra work which is unnecessary and will add time and money.

Functional skills

Organising work in sequence will require you to understand all the tasks that need to be done. To do this you will be practicing FE 1.2.3 – Read different texts and take appropriate action. To draw up your own plan for work you will practice FE 1.3.1 – 1.3.5 Write clearly with a level of detail to suit the purpose. You will also cover FM 1.3.1 Judge whether findings answer the original problem and FM 1.3.2 Communicate solutions to answer practical problems.

Experience helps when planning a sequence of operations. There are times when mistakes are made in the sequencing and things have to be re-done, but it is always a good idea to list the sequence of operations before starting the work.

Woodworking joints

At the end of this section you should be able to:

- understand simple jointing methods used on doors and windows
- identify the main joints used in the assembly of units and fitments
- state the correct jointing methods used on a common staircase.

Halving joints

Halving joints are used where two pieces of timber overlap and can be used for lengthening or on corners. They are relatively simple joints to construct, but variations can make them stronger and more difficult such as dovetail or mitred halving.

Edge joints

Edge joints are where the timber is joined along its edge and is used in lengthening, usually in the form of a finger-type joint.

Lengthening

Other edge-type joints are simple butt joints where a decorative piece of timber is placed along the edge of a material such as plywood to disguise the rougher edge. In some cases the edge joint can take the form of a tongue and groove joint.

Draw boring

To give additional strength to a through mortise and tenon style joint and to pull the joint tight, a timber dowel is fixed through the face of the frame's head and into the tenon. The hole should be previously drilled in the face and then slightly off-centre in the tenon. When the dowel is knocked home the joint is pulled tight. This method is known as draw boring.

Joints used on doors and windows

During the manufacture of doors and windows the mortise and tenon joint is extensively used. The type of mortise and tenon will depend on its location. Examples of this joint are described over the next few pages.

Through mortise and tenon

In a through mortise and tenon joint a single rectangular tenon is slotted into a mortise. See Figure 34.22.

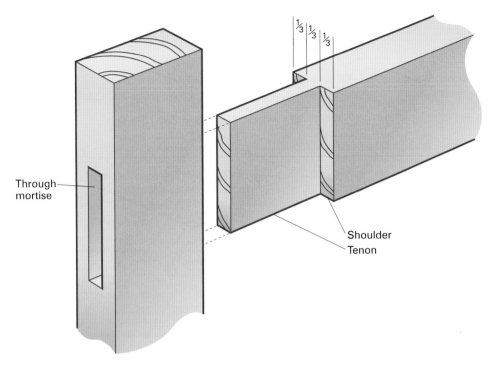

Figure 34.22 Through mortise and tenon

Stub mortise and tenon

In a stub mortise and tenon joint the tenon is stopped short to prevent it protruding through the member. See Figure 34.23.

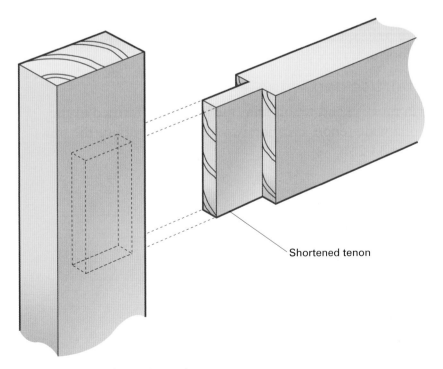

Figure 34.23 Stub mortise and tenon

Haunched mortise and tenon

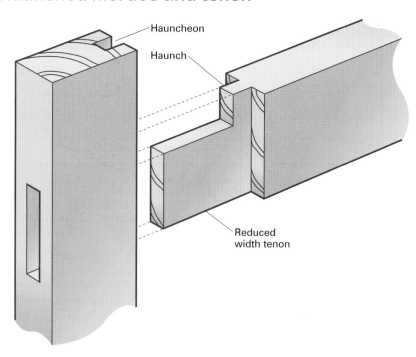

Figure 34.24 Haunched mortise and tenon

In a haunched mortise and tenon joint the tenon is reduced in width, leaving a shortened portion of the tenon protruding which is referred to as a haunch. See Figure 34.24. The purpose of the haunch is to keep the tenon the full width of the timber at the top third of the joint. This will prevent twisting. A haunch at the end of the member will aid the wedging-up process and prevent the tenon becoming **bridled**. For a detailed description of the wedging-up process look at Unit 2036.

Twin mortise and tenon

In a twin mortise and tenon joint the haunch is formed in the centre of a wide tenon, creating two tenons, one above the other. See Figure 34.25.

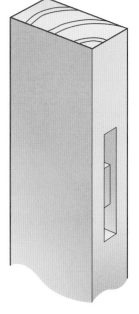

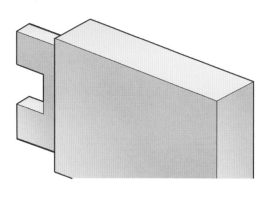

Figure 34.25 Twin mortise and tenon

Double mortise and tenon

For a double mortise and tenon, two tenons are formed within the thickness of the timber. See Figure 34.26.

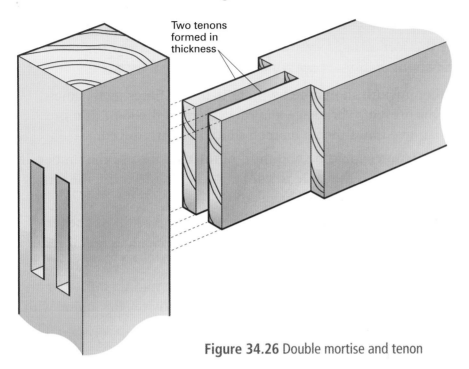

Two tenons formed in thickness

Figure 34.26 Double mortise and tenon

Stepped shoulder joint

Used on frames with rebates, a stepped shoulder joint has a shoulder stepped the depth of the rebate. This joint can also be combined with haunched, twin or stub tenons. See Figure 34.27.

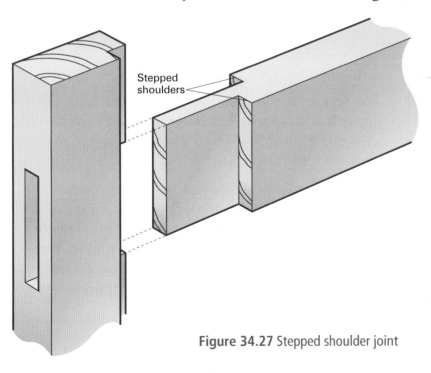

Stepped shoulders

Figure 34.27 Stepped shoulder joint

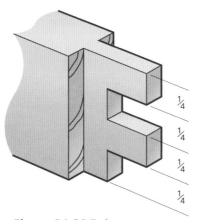

Figure 34.28 Twin tenon with twin haunch

Twin tenon with twin haunch

A twin tenon with twin haunch joint is used on the deep bottom rails of doors. See Figure 34.28.

Basic rules on mortise and tenon joints

The proportions of mortise and tenon joints are very important to their strength. Some basic rules are as follows:

- Tenon width should be no more than five times its thickness. This prevents shrinkage and movement in the joint. If more than five times then a haunch should be introduced.

- The tenon should be one-third of the thickness of the timber. If a chisel is not available to cut a mortise at one-third, the tenon should be adjusted to the nearest chisel size.

- When a haunch is being used to reduce the width of a tenon, then about one-third of the overall width should be removed. The depth of a haunch should be the same as its thickness.

- Although a tenon should be located in the middle third of a member, it can be moved either way slightly to stay in line with a rebate or groove.

Joints used in units and fitments

During the design and setting out of units and fitments the most common joint used is the mortise and tenon, but these are not the best if there are forces likely to try to pull the joint apart. These are called **tensile forces**.

Parts of a unit or fitment subject to such forces must incorporate a joint design that will allow for this. A drawer on a unit is often subject to tensile forces, so a dovetail joint would be used.

Key term

Tensile forces – a force that is trying to pull something apart

Dovetail joints

The two most common types of dovetail joint are through and lapped. A through dovetail joint is shown in Figure 34.29 and a lapped dovetail in Figure 34.30.

Dovetail joints should have a slope (sometimes called the pitch) of 1:6 for softwoods, or 1:8 for hardwoods. If the slope of the dovetail is excessive then the joint will be weak due to short grain. If the slope is insufficient the dovetail will have a tendency to pull apart. The slope (or pitch) is shown in Figure 34.31.

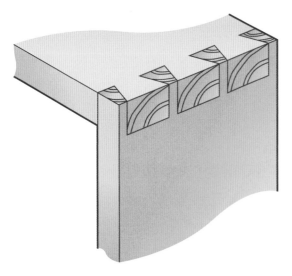

Figure 34.29 Through dovetail joint

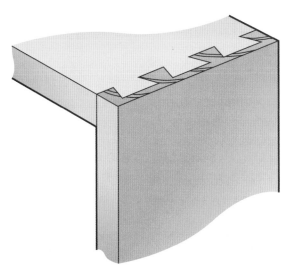

Figure 34.30 Lapped dovetail joint

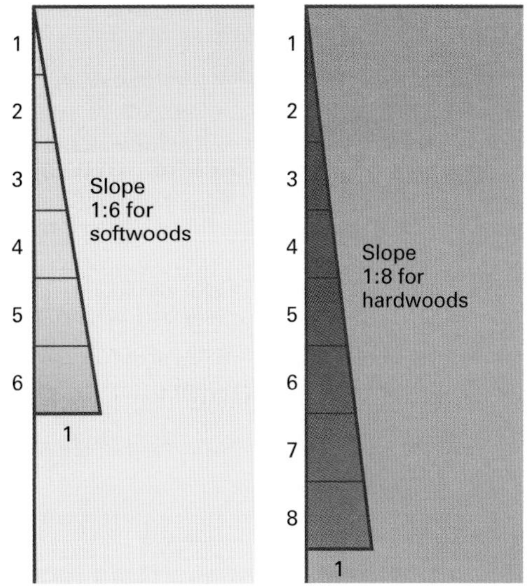

Slope
1:6 for
softwoods

Slope
1:8 for
hardwoods

Figure 34.31 Slope of a dovetail joint

Correct jointing methods on staircases

The most common joint used in staircase construction is a **stopped housing joint**. This joint is used to locate or house the tread and riser of a step into the string. It will be stopped at the nosing of the tread. The minimum housing depth is 12 mm. See Figure 34.32.

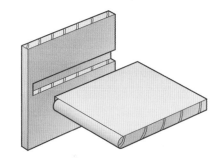

Figure 34.32 Stopped housing joint (staircase)

When the string of a stair meets a newel post, a stubbed and haunched mortise and tenon joint is used, as shown in Figure 34.33. More information on this type of joint can be found earlier in this unit.

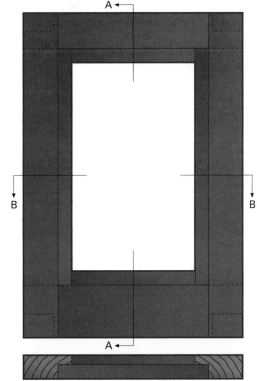

Vertical section A/A

Horizontal section B/B

Figure 34.33 Housed string with mortise and tenon

> **Key term**
>
> **Stopped housing joint** – where a cut does not go completely through the timber

> **Did you know?**
>
> The best way to understand how joints are assembled and used is to get some timber, some tools, get in a workshop and practise, practise, practise!

FAQ

How do I choose between hardwood and softwood for a job?

The type of timber you should use for a job is usually detailed in the specifications. There are several reasons why one type of wood is chosen over another. Hardwood is usually more expensive and longer-lasting than softwood. It is often grown in the hot climates of equatorial countries (e.g. African and South American countries) and is used for jobs where the wood will be visible (i.e. high-class joinery). Softwood is usually grown in countries with cooler climates. It is often cheaper than hardwood and used for jobs where the wood will be concealed (e.g. floorboards and rafters).

Why would you need sawn sizes on a cutting list?

By putting sawn sizes on the cutting list the machinist will quickly be able to determine the most cost-effective sections of stock to use from the timber rack.

Check it out

1   What drawings should setting out rods contain?
2   Why should rods be stored after the component is made?
3   Explain the purpose of face and edge marks.
4   Explain the process of marking out.
5   Describe the purpose of cutting lists.
6   Explain what a haunch is and why is it used.
7   What type of joint is used in door frames?
8   Explain what draw boring is.
9   Show, using sketches, a way in which edge joints can be formed.
10 Show, by way of a sketch, three different halving joints.
11 Explain the importance of deadlines.
12 Explain the purpose of a sequence of operations.

Getting ready for assessment

The information contained in this unit, as well as continued practical assignments that you will carry out in your college or training centre, will help you with preparing for both your end of unit test and the diploma multiple-choice test. It will also aid you in preparing for the work that is required for the synoptic practical assignments.

The information contained within this unit will aid you in learning how to identify and calculate the materials and equipment required to produce setting out details for routine joinery products.

You will need to be familiar with:

• interpreting information for setting out
• selecting resources for setting out
• setting out for bench joinery and site carpentry.

Setting out is a key stage in working in carpentry. Learning outcome three has shown the correct tools that are needed for setting out and explained the uses and proportions of different joints. You will need to use this knowledge to measure and mark out using the correct tools and to specifications. There is a large range of items that need to be set out in carpentry and joinery, such as doors, frames, linings, units, staircases, studwork, floor joists and wall plates.

This unit has also explained the importance of deadlines and organising procedures into the correct sequence. This will be vital for both the practical assignments you will carry out and in your professional life. Meeting deadlines is important for your reputation and procedures will help you to do this without compromising the quality of your final work.

Before you start work on the synoptic practical test it is important that you have had sufficient practice and that you feel that you are capable of passing. It is best to have a plan of action and a work method that will help you. You will also need a copy of the required standards, any associated drawings and sufficient tools and materials. It is also wise to check your work at regular intervals. This will help you to be sure that you are working correctly and help you to avoid problems developing as you work.

Your speed at carrying out these tasks will also help you to prepare for the time limit that the synoptic practical task has. But remember, don't try to rush the job as speed will come with practice and it is important that you get the quality of workmanship right.

Always make sure that you are working safely throughout the test. Make sure you are working to all the safety requirements given throughout the test and wear all appropriate personal protective equipment. When using tools, make sure you are using them correctly and safely.

Good luck!

CHECK YOUR KNOWLEDGE

1 Plans or drawings should be set out, full scale on a:
  a marking out rod.
  b setting out rod.
  c setting out board.
  d marking out board.

2 Why are setting out rods often painted white?
  a so they can be easily found.
  b to conceal any defects in the wood.
  c to aid the clarity of drawing.
  d to protect the wood.

3 What can also be marked on setting out rods?
  a general remarks.
  b the date the setting out rod was created.
  c rough sketches.
  d elevations.

4 Hardwood trees are:
  a coniferous.
  b evergreen.
  c deciduous.
  d pine trees.

5 The difference between hardwood and softwood is:
  a hardwoods are hard.
  b hardwoods are soft.
  c there is no difference.
  d botanical.

6 A timber that is creamy white with occasional dark streaks, and can be used for building furniture and boats, is:
  a pitch pine.
  b teak.
  c ash.
  d western red cedar.

7 A man-made board with compressed wood flakes is called:
  a plywood.
  b blockboard.
  c chipboard.
  d fibreboard.

8 Which tool could you use instead of a pencil, when you want to mark measurements very accurately?
  a marking gauge.
  b tri-square.
  c marking knife.
  d callipers.

9 What are dividers used for?
  a accurate checking of widths and gaps.
  b marking out measurements.
  c measuring angles.
  d marking length and thickness.

10 Which tool is best for marking deep lines?
  a mortise gauge.
  b marking gauge.
  c cutting gauge.
  d combination square.

Know how to mark out from setting out details for routine joinery products

Although marking out and setting out are often taught together, they are in fact two separate – but related – skills. Where setting out can be compared with drawing a route map to the finished piece – creating the detailed drawings that will help timber become object, marking out is the skill needed to transfer the detailed drawings onto the timber itself. Much like an artist wouldn't put paintbrush to canvas without first plotting the picture he or she is about to create, nor would a good carpenter or joiner begin a project simply by cutting or planing wood at random.

This unit also supports NVQ Unit VR15W Mark Out from Setting Our Details for Routine Joinery Products.

This unit also contains material that supports TAP Unit 6 Mark Out from Setting out Details.

This unit will cover the following learning outcomes:

- How to produce marking out effectively
- How to produce accurate marking out.

K1. Know how to produce marking out effectively

Effective marking out is about more than just transferring marks to timber. A good carpenter or joiner will know which faces of the timber should be marked out and will transfer his or her instructions clearly. An excess of information on the rod will lead to confusion. At best, this will cause small delays as questions are asked and clear instructions confirmed. At worst, entire jobs may need to be re-done if information is misinterpreted or not understood.

Basic marking out

Marking out is the transfer of the information on a setting out rod to the timber. It is a very important process and should be checked thoroughly. Wrong information transferred to the timber at this stage will result in errors during assembly. Such errors are likely to be time consuming and costly.

At the end of this section you will be able to:

- select the correct sides of timber on which to mark out; these are known as the face and edge
- transfer information from setting out rod to timber.

Face and edge marks

The face and edge are the two most important of the four sides of a piece of timber. They are, therefore, usually selected as the two best adjacent sides.

The position and severity of any defects present in the timber should also be noted at this point. If these can be removed when cutting rebates or grooves then the sides containing them may still be the best sides to use as the face and edge.

Face and edge marks are clearly applied to the relevant sides after careful inspection of the timber. They are used as a reference point from which all marking out is completed.

Items such as door stiles must be marked out in pairs, since they are not reversible. In this case the face and edge marks must always be opposite to each other.

Face side and edge are normally the front and inside edges of the framework. However, where one side of the frame is not flush with the other, such as in a casement window and sill, then the flush side should be chosen as the face.

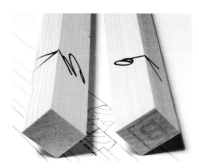

Figure 35.1 Paired members

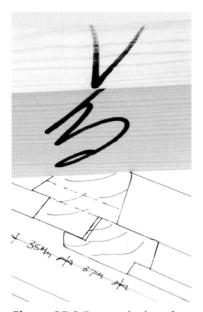

Figure 35.2 Face and edge of timber

Transferring information from a setting out rod

Marking out on the timber should be as clear and simple as possible with no unnecessary lines, as these cause confusion and possible errors. Pencil lines should be clear, sharp and made with a hard pencil. Wherever possible, all marking out should be completed in a single operation and any wrongly placed or double lines should be rectified immediately.

The importance of deadlines when marking out

The importance of deadlines is simple – the job must be done on time, otherwise things can go horribly wrong. Delays when marking out can have massive knock-on effects. For example, if a client wants a window fitted on the 20th and the window isn't made in time, then the workers who would be fitting the window will have no work for that day but will still need to be paid. The client will not have the job done as scheduled and they may have arranged time off from work. That will affect the company's image with regard to future jobs. The window will need to be fitted on a different day, adding to the overall cost of the job.

On larger jobs a penalty clause may be imposed that states if the work is not completed by a certain date, the company will be fined by the client. This can be very costly, as some contracts have clauses that cost the company thousands of pounds for each day over the target date that work continues.

Any delays or errors that marking out has caused will affect more than just the current part of the job or even the whole job. Things will have to be re-done correctly, which will take time. The added time will delay completion of the first job and could roll over causing delays to the next job.

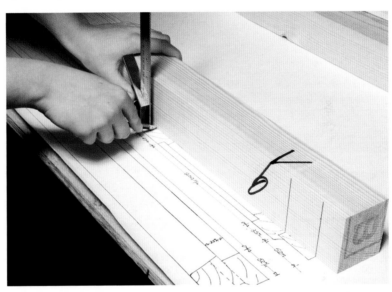

Figure 35.3 A piece of timber being marked out from a door rod

Figure 35.4 The shoulder and mortise lines are directly transferred and a completed section of timber is drawn across the member

K2. Know how to produce accurate marking out

Accurate marking out is about more than transferring marks onto timber. In order to do the job to the high standard required, a good carpenter or joiner will be familiar with a range of site drawings and documents, all of which will be useful when it is time to begin marking out. Schedules are important when it comes to planning any type of work, but are particularly useful when it comes to marking out. If you can see on your schedule that the doors are due to be hung in March, while window frames aren't scheduled till April, you will know to complete the door work before starting on the windows. Building regulations, specifications, cutting lists, plans and many other types of on site documentation will be vital to your work in marking out. They have been covered exhaustively in Unit 2002.

Marking out basic components

While marking out is a skill that applies to all manner of carpentry and joinery products, there are some more routine components you will need to make more often than others. After all, every house has windows and doors, and most of them have stairs! This section covers those components you will need to mark out most often.

Windows

As windows are expected to stand up to the elements, they usually need to be made from a solid construction.

In this section, our example is a simple casement window, 900 mm high and 500 mm wide.

Setting and marking out a window

First, make setting out rods drawn at full size to show the sections of the height and width.

Use this drawing to produce the cutting list, then use the cutting list to prepare the materials for marking out.

When manufacturing a window, two framing methods are used: one for the fixed parts and another for the moving parts. In our example, the fixed part is the window frame, and the moving part is the opening sash. For the frame, the horizontal members (head, sill) are mortised and the vertical members (jambs) are tenoned. For the moving sash, the opposite is true – the horizontal members are tenoned and the vertical members mortised.

> **Did you know?**
>
> The method of setting out and marking is virtually identical for every component. Remember this as you look at the following examples of manufactured components

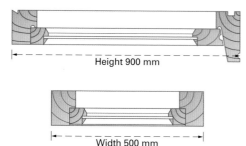

Height 900 mm

Width 500 mm

Figure 35.5 Height and width rods

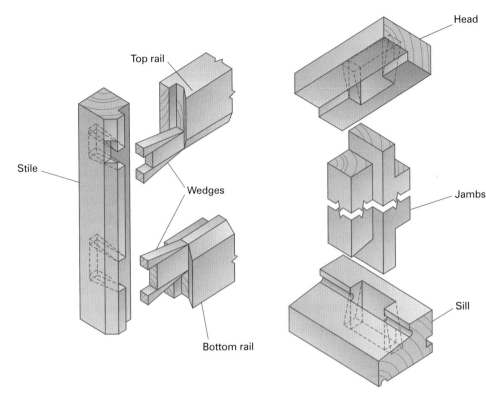

Stile

Top rail

Wedges

Bottom rail

Sash construction

Head

Jambs

Sill

Frame construction

Figure 35.6 Joints in frames and sashes

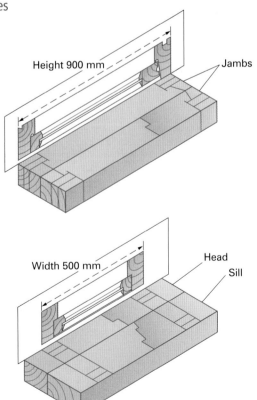

Height 900 mm

Jambs

Width 500 mm

Head

Sill

Once the materials are machined, you can mark them out by transferring marks from the setting out rod to the various members.

Once all members are marked out, the joints, rebates, grooves, etc. can be machined ready for assembly.

Figure 35.7 Members marked out

Doors

While there are many types of doors, they fit into one of two categories:

- **flush/hollow core door** – these are hollow doors with a frame around the outside, clad usually with hardboard or plywood. For strength, the interior of the door is packed with cardboard and a lock block is fitted to one of the stiles to allow a lock or latch to be fitted

- **framed door** – these are made from hardwood or softwood and constructed using either mortise and tenon or dowel joints. Once made, the frame is usually rebated to incorporate solid wood panels or glass, onto which beads will be fitted.

As flush doors and dowel-construction framed doors are normally factory-produced, here we will look at what is involved in the production of a mortise and tenon framed door.

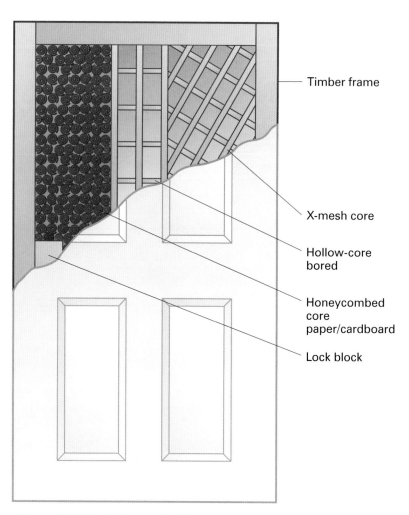

Figure 35.8 Flush door exploded view

Timber frame

X-mesh core

Hollow-core bored

Honeycombed core paper/cardboard

Lock block

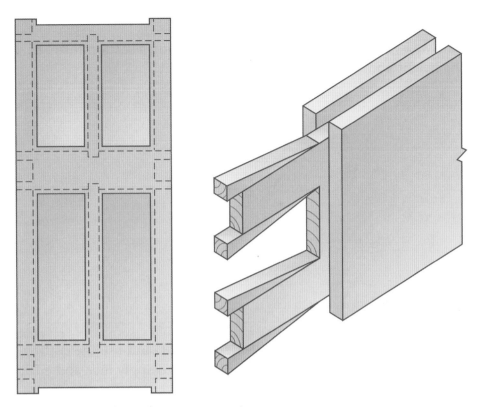

Figure 35.9 Mortise and tenon construction

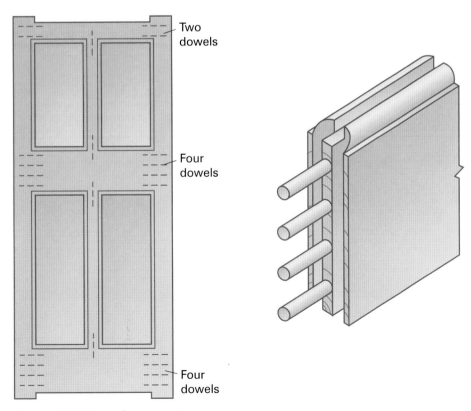

Two
dowels

Four
dowels

Four
dowels

Figure 35.10 Dowel-construction

Setting out and marking a door

As always, begin with setting out rods showing a section through the height and width.

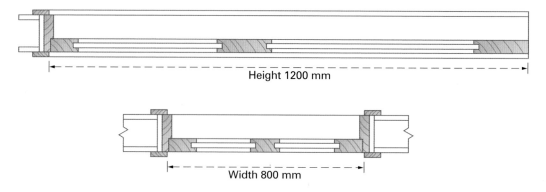

Figure 35.11 Height and width rods

You can use the rods to produce a cutting list, and use the cutting list to machine the members, remembering to apply face and edge markings. Next, transfer the marks from the rods to mark out all the members, remembering to mark out the horizontal members together and the vertical members together.

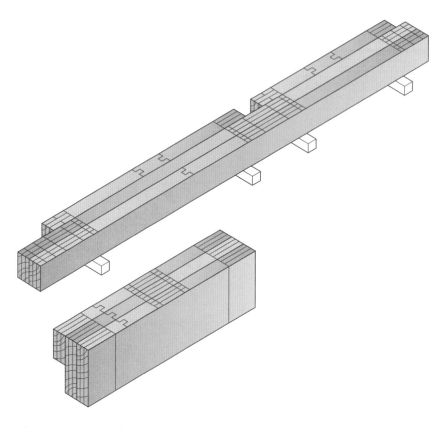

Figure 35.12 Members marked out

Stairs

Stairs are set out and marked out differently from doors and windows. To set out stairs, you first need all the dimensions, including the rise, going, etc.

Two templates are needed to mark out the tread and riser positions on the strings – the pitch board and the tread and riser template.

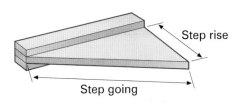

Figure 35.13 Pitch board

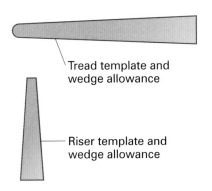

Figure 35.14 Tread and riser template

Setting out and marking stairs

First, machine the strings to the required sizes and set them out on the bench, remembering to make the face and edge marks.

Mark the pitch line using a margin template for accuracy.

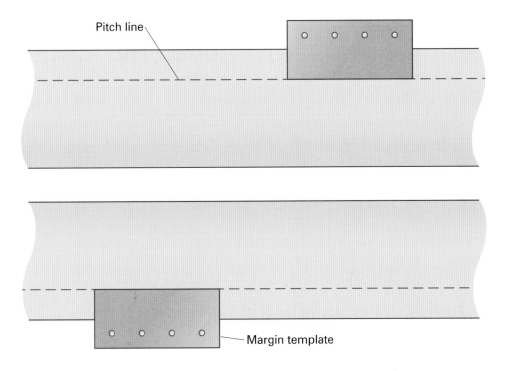

Figure 35.15 Marking the pitch line using a margin template

Set a pair of dividers to the hypotenuse of the pitch board, then mark this distance all the way along both strings. The two points of intersection will establish the tread and riser points relevant to the pitch line.

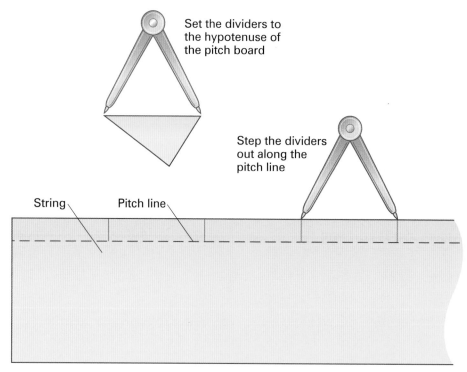

Set the dividers to the hypotenuse of the pitch board

Step the dividers out along the pitch line

String

Pitch line

Figure 35.16 Marking the tread and riser points

The pitch board can now be used to mark the rise and going onto the strings. The riser and tread templates can be used to mark out the position of each step.

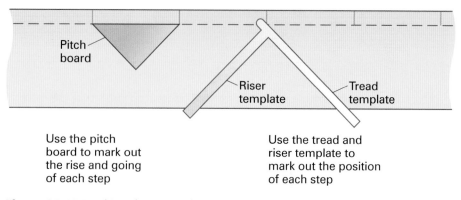

Pitch board

Riser template

Tread template

Use the pitch board to mark out the rise and going of each step

Use the tread and riser template to mark out the position of each step

Figure 35.17 Marking the rise and going and the position of the treads

Figure 35.18 Router stair-housing jig

Now router out the housings. It is best to use a stair-housing jig combined with a router for this job.

Figure 35.19 Stair-housing jig used with router

The treads and risers can now be machined. The treads usually have a curved nosing and a groove on the underside, to allow the risers to be housed into the treads.

You are now ready to assemble the stairs.

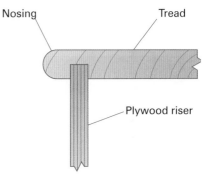

Figure 35.20 Joint between tread and riser

Working life

Marking out mix-up

Molly, a second-year apprentice, has been given the task of manufacturing a number of replacement sashes for a housing re-development project in the local area. All setting out rods were drawn up by Phil, a recently qualified bench joiner. Molly was given all the rods with a range of component drawings referenced to the specific rod. After marking out the first sash, Molly realised that the size written on the drawings did not match the size on the rod as drawn, but did match another sash to be made.

What do you think has happened?

What action should Molly take to overcome this problem?

What should be done to prevent this from happening again?

Functional skills

Resolving problems in your work will give you the opportunity to practice the Interpreting elements of functional skills, e.g. FM 1.3.1 Judge whether findings answer the original problem and FM 1.3.2 Communicate solutions to answer practical problems.

FAQ

When marking out for repetitive items of joinery, for example a large number of standard-size doors, would you need to mark each piece of stock from the rod separately?

No, if you transfer the information to one piece of stock, you can clamp all the pieces that will be the same size together. From there you can transfer joint lines to all pieces. You must remember to ensure that all pieces are in pairs using your face and edge marks.

Check it out

1   State the purpose of mullions and transoms.
2   Explain why the dry assembly stage is important for a window.
3   Explain what a winding is and how can you check for it.
4   Explain the purpose of setting out.
5   Explain the following:
   a sight size
   b shoulder size
   c overall size.
6   State the best way to cut out the housing on a stair string.
7   Explain when a tenon with a stepped shoulder should be used.
8   A door stile has a finished section of 95 mm x 44 mm. What size chisel would be used to cut the mortise?

Getting ready for assessment

The information contained in this unit, as well as continued practical assignments that you will carry out in your college or training centre, will help you with preparing for both your end of unit test and the diploma multiple-choice test. It will also aid you in preparing for the work that is required for the synoptic practical assignments.

The information contained within this unit will aid you in learning how to identify and calculate the materials and equipment required for marking out, from setting out, details for routine joinery products.

You will need to be familiar with:

- how to produce marking out efficiently
- how to produce accurate marking out.

Marking out involves transferring the information you have set out on drawings onto the timber itself. You will need to remember the skills you learnt in Unit 2034 in order to complete the work accurately. For learning outcome one you have seen how delays can have an effect on the overall programme of work, in terms of waste of material and slowing of work programme. This should show you the importance of producing making out to timescales. It is also important that you produce marking out with minimal errors. You will need to produce detailed cutting lists, that takes into account the information contained in the drawings.

Before you start work on the synoptic practical test it is important that you have had sufficient practice and that you feel that you are capable of passing. It is best to have a plan of action and a work method that will help you. You will also need a copy of the required standards, any associated drawings and sufficient tools and materials. It is also wise to check your work at regular intervals. This will help you to be sure that you are working correctly and help you to avoid problems developing as you work.

Your speed at carrying out these tasks will also help you to prepare for the time limit that the synoptic practical task has. But remember, don't try to rush the job as speed will come with practice and it is important that you get the quality of workmanship right.

Always make sure that you are working safely throughout the test. Make sure you are working to all the safety requirements given throughout the test and wear all appropriate personal protective equipment. When using tools, make sure you are using them correctly and safely.

Good luck!

CHECK YOUR KNOWLEDGE

1 Face and edge marks are placed on:
   a the best face and edge.
   b the worst face and edge.
   c any face and edge.
   d all faces and edges.

2 What is sawn size?
   a the size of the timber prior to cutting and planing.
   b the size of the timber after cutting and planing.
   c the size of the saw.
   d the size of the saw teeth.

3 The best way to mark out timber for a door is to:
   a measure each piece individually.
   b transfer the measurements from a setting out rod.
   c guess the measurements.
   d measure from face marks.

4 How many templates are used in the marking out of a stair string?
   a 1
   b 2
   c 3
   d 4

5 Which joint is used to fit the top of the risers to the underside of the treads?
   a housing joint.
   b butt joint.
   c tenon joint.
   d dovetail joint.

6 The majority of modern windows are:
   a bay windows.
   b dormer windows.
   c casement windows.
   d box sash windows.

7 Which power tool is used to cut out the housings on a stair string?
   a jig saw.
   b planer.
   c router.
   d drill.

8 A door that contains a lock block is a:
   a framed door.
   b flush door.
   c fire door.
   d double door.

9 What is the name for a door which contains glass panels?
   a glossed.
   b glass panelled.
   c glazed.
   d glass plated.

UNIT 2036

Know how to manufacture routine joinery products

The manufacture of routine joinery products is a key part of any carpenter's skill set. Throughout this book we have looked at the importance of selecting and maintaining the right tools and equipment for the job. This is as important in joinery as it is in all other aspects of woodwork. Selecting the wrong timber, or using the wrong tools for a job, will result in poor quality work, and will lead to schedule delays and added costs when work needs to be re-done.

Knowing how to manufacture routine joinery products will give you more control over a project. Even if you do not do the manufacturing yourself, knowing what the work entails will mean you will be in a better position to estimate schedules and the quantities and resources needed.

This unit also supports NVQ Unit VR16B Manufacture Routine Products.

This unit also contains material that supports TAP Unit 7 Manufacture Joinery Products.

This unit will cover the following learning outcomes:

- How to select correct materials
- How to manufacture joinery.

K1. Know how to select correct materials

Timber selection is a key aspect of all carpentry and joinery jobs. Using the wrong type of wood for the job can lead to unintended results. Unit 2008 covers the identification and selection of timber on pages 132–135. Tables of common hardwoods and softwoods can be found on pages 133–135.

There are a number of other materials to select before a job can be completed. These include ironmongery, fixtures and fittings, and adhesives, among others. Information on materials storage can be found on page 22–25.

Materials storage, as well as being important for safety, is also important in efficiency. Storing materials near to where they will be used prevents double handling issues and allows the job to be completed more quickly.

Ironmongery

Ironmongery includes fittings and fixtures made of iron, as well as hardware made from other materials including brass, chrome, porcelain and glass. Items you might come across include door handles, hooks, locks, hinges, window fittings, screws and bolts.

Door furniture

Door furniture such as locks, bolts, letter boxes, knockers and handles etc. are 'desirable' items, which means that they are very likely to be stolen unless stored securely. A store person is usually responsible for the storage and distribution of such items and keeps a check on how many are given out and to whom.

In addition to being stored securely, door furniture should also be kept in separate compartments of a racking storage system, or at least on shelving. Where possible, all door furniture should be retained in the manufacturer's packaging until needed. This prevents damage and loss of components such as screws and keys. Large heavy items should be stored on lower shelves to avoid unnecessary lifting.

Ironmongery for external doors has been covered in Unit 2009, page 185–188.

Figure 36.1 Examples of door furniture

Fixings

Each type of fixing is designed for a specific purpose and includes items such as nails, screws, pins, bolts, washers, rivets and plugs. As with door furniture, fixings tend to disappear if their storage is not supervised and controlled by a store person.

Fixings must be stored appropriately to keep them in good condition and to make them easy to find. Where possible, they should be kept in bags or boxes clearly marked with their size and type. Storing them in separate compartments is the most convenient method of storage. This enables easy and fast selection when required and prevents the wrong fixing being used. Time taken to sort out different types and sizes of fixings that have become mixed up is a waste of your time and your employer's time.

Adhesives

Adhesives are substances used to bond (stick) surfaces together. Because of their chemical nature, there are a number of potentially serious risks connected with adhesives if they are not stored, used and handled correctly.

All adhesives should be stored and used in line with the manufacturer's instructions. This usually involves storing them on shelving, with labels facing outward, in a safe, secure area (preferably a lockable store room). It is important to keep the labels facing outwards so that the correct adhesive can be selected.

The level of risk associated with adhesive use is dependent on the type of adhesive. Some of the risks include:

- explosion
- poisoning
- skin irritation
- disease.

As explained in Unit 1001, these types of material are closely controlled by COSHH, which aims to minimise the risks involved with their storage and use.

All adhesives have a recommended **shelf life**. This must be taken into account when storing to ensure the oldest stock is stored at the front and used first. Remember to refer to the manufacturer's guidelines as to how long the adhesive will remain fit for purpose once opened. Adhesives can be negatively affected by poor storage, including loss in adhesive strength and extended setting time.

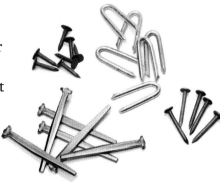

Figure 36.2 Different types of fixings should be stored separately

Key term

Shelf life – how long something will remain fit for its purpose while being stored

Figure 36.3 Adhesives should be stored according to the manufacturer's instructions

Unit 2036 Know how to manufacture routine joinery products

Tool selection

In carpentry and joinery, it is vital to select the right tools for the job. Tool maintenance is key – poorly maintained saw teeth, for example, can ruin a beautiful piece of wood. Ensuring your tools are in good condition is your responsibility. If a job is delayed because you've had to re-work some timber, you will have added to the schedule length and increased the total cost of the job.

The importance of safely storing your tools, and of the need for regular maintenance checks, has been covered exhaustively in Unit 1001, pages 22–25.

K2. Know how to manufacture joinery

Once a bench joiner has produced the setting out details for the specified task and transferred the information onto the timber stock to be used, a joiner can begin the task of cutting and assembling the required item of joinery.

Frames and linings

Window and door frames / linings are fitted into openings left in masonry and hold the window or door in place. It is very important to make sure the frame or lining fits the opening well and that it is level. If this is not done, you will end up with doors and windows that don't hang properly and don't open and close properly.

Door frames and linings

Door frames are usually of a substantial and solid construction. They are mortised and tenoned and comprise heads, jambs and sills. They usually have a rebate cut from the solid timber. See Figure 36.4.

Door linings are of much lighter construction than frames and are used exclusively for internal doors. They don't usually have sills and are normally jointed between head and jambs with a form of housing joint, such as tongued housing (see Figure 36.5). Door linings usually have planted (nailed on) stops.

Door frames and linings are usually assembled in the joiner's shop using a tongued housing joint (see Unit 2034). To give additional strength and to pull the joint tight, a timber **dowel** is fixed through the face of the frame's head and into the tenon. The hole should be previously drilled in the face and then slightly off-centre in the tenon. When the dowel is knocked home the joint is pulled tight. This method is known as draw boring. The hole needs to be offset as shown in Figure 36.7.

> **Key term**
>
> **Dowel** – a headless wood or metal pin used to join timber

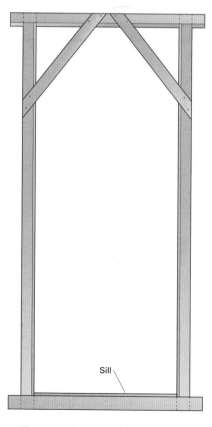

Sill

Figure 36.4 Door frame

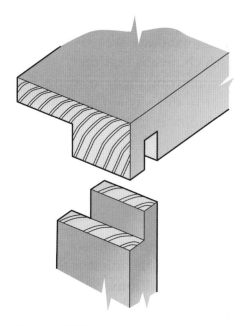

Figure 36.5 Tongued housing joint

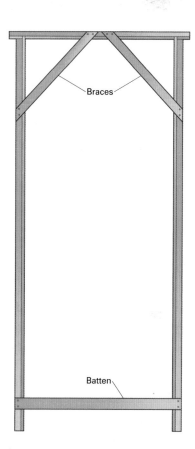

Figure 36.6 Door lining

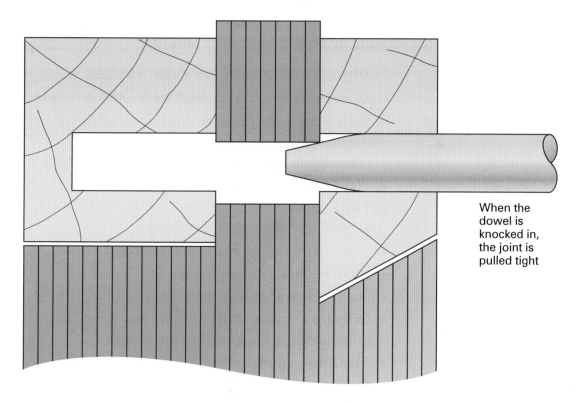

When the
dowel is
knocked in,
the joint is
pulled tight

Figure 36.7 Draw boring and pin

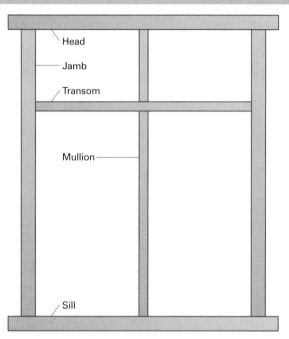

Figure 36.8 Window frame

Key terms

Mullions– vertical dividers in a window frame

Transoms– horizontal dividers in a window frame

Stiles – the vertical members of the sash; the hinges are fitted on to one of them

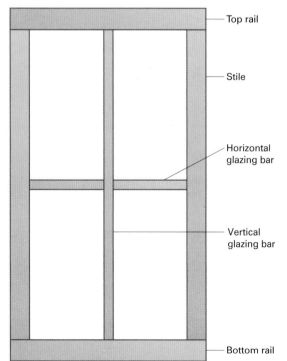

Figure 36.9 An opening casement

Window frames

The majority of modern windows are casement windows, which means that they are hinged on the side allowing them to swing open vertically (similar to the way a door opens). A casement window is made up of two main components:

- the frame
- the opening casement.

The frame

Like a door frame, a window frame consists of a head, sill and jambs. When the frame is to be divided, members called **mullions** (vertical dividers) and **transoms** (horizontal dividers) are included (see Figure 36.8).

The opening casement

The opening part of the window, known as the sash, consists of a top rail, bottom rail and two **stiles**. When the opening of the casement sash is to be divided up further, glazing bars are used (see Figure 36.9). The procedure for hanging a casement sash is exactly the same as for a door, only usually on a smaller scale. The procedure is summarised below, but look back at Unit 2009 for the full door hanging procedure (page 174–175).

Hanging casement window sashes

- Mark the hanging side on both the frame and the sash.
- Cut off any horns (these are waste stock left overhanging on the stiles to aid cleaning up and prevent damage to sash corners prior to installation).
- Plane to fit the hanging stile.
- Plane the sash to the required width, running parallel with the side of the frame.
- Plane to fit the top and bottom of the frame.
- Mark out and cut the hinges.
- Screw one leaf of each hinge to the sash.
- Offer up the sash to the opening and screw the other leaves of the hinges to the frame.
- Make fine adjustments if needed and fit specified ironmongery.

General assembly procedure

There are some general points you should be aware of when assembling a door or window frame:

- dry assembly
- squaring up
- checking for winding
- wedging up.

We will look at each of these in turn over the next few pages.

Dry assembly

All the timber making up the frame should be knocked together dry prior to final assembly. 'Dry', in this situation, doesn't mean in dry conditions (i.e. out of the rain), but rather without the use of adhesive (i.e. a dry or practice run). This ensures that all joints are a good fit and that the frame is:

- the correct size
- square
- not **winding**.

Squaring up

When the frame has been assembled, glued and cramped up (held together whilst drying with clamps – see Figure 36.10), it should be tested to make sure that it is square, meaning that the corners are at right angles. The most accurate way of doing this is to compare the diagonals using a squaring rod, which is a piece of rectangular timber with a small nail or panel pin knocked into the end (see Figure 36.11). The protruding nail is placed in one corner of the frame and the corner of the opposite diagonal is marked on the rod with a pencil. This procedure is repeated for the two other corners of the frame. If both pencil lines match up, the frame is square.

If the pencil lines do not match up, the frame needs to be adjusted by angling the cramps and pushing the frame. This should be done until the two pencil marks match up, meaning that the diagonals are the same length and the frame is square.

> **Key term**
>
> **Winding** – twisted (wood), as when a wood frame is twisted or skewed

> **Remember**
>
> Once the frame is assembled dry, check the sizes one more time. As soon as the frame is glued together it will be too late to adjust anything

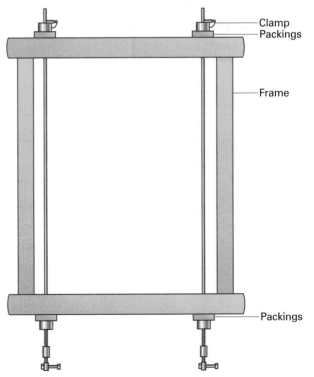

Figure 36.10 Cramping up a frame

Labels: Clamp, Packings, Frame, Packings

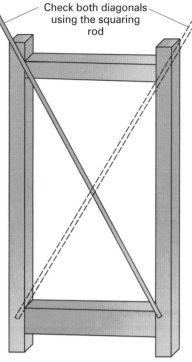

Check both diagonals using the squaring rod

Figure 36.11 Squaring up with a squaring rod

> **Remember**
>
> When you have made any adjustments to remove winding, go back and make sure the frame is still square

Checking for winding

Winding is a term that describes a frame that is twisted. You will need to check that the frame you have assembled is not twisted by using winding rods. These are simply two pieces of timber which are laid parallel across the frame when it is lying flat on the workbench. Close one eye and look across the winding rods. They should be parallel. If they are, the frame has no twist. If the winding rods are not parallel, the frame is winding and adjustments will have to be made to the joints before further assembly is carried out.

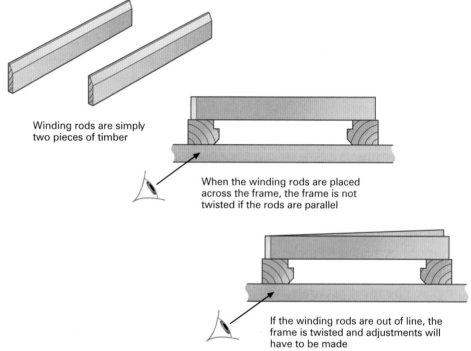

Winding rods are simply two pieces of timber

When the winding rods are placed across the frame, the frame is not twisted if the rods are parallel

If the winding rods are out of line, the frame is twisted and adjustments will have to be made

Figure 36.12 Checking for winding

Wedging up

The haunched mortise and tenon joints you have used in the assembly of your door or window frame, are not only held in place by glue, but they are also wedged. This should be done after the frame has been glued, squared and checked for winding. Wedging up involves placing a small wedge on either side of the tenon and carefully driving it in to ensure a good tight joint. To help keep the frame square and the joints pulled in tight, it is best to drive in the external wedge (the haunch side) first. Look back at Unit 2034 to remind yourself about haunched mortise and tenon joints.

Doors

There are a number of different types of door assembled by carpenters and joiners. In this section we will only look at panelled and glazed doors, since they incorporate all the principles that you need to know about to be able to assemble any door.

Panelled doors

Panelled doors have a frame made from solid timber rails and stiles (see Figure 36.14). When made by the bench joiner in the workshop, they will almost certainly incorporate a mortise and tenon type of joint. The frame will either be grooved or rebated to receive a number of either plywood or timber panels.

Glazed doors

Glazed doors are made in a similar fashion to panelled doors, however one or more of the panels is replaced by glass.

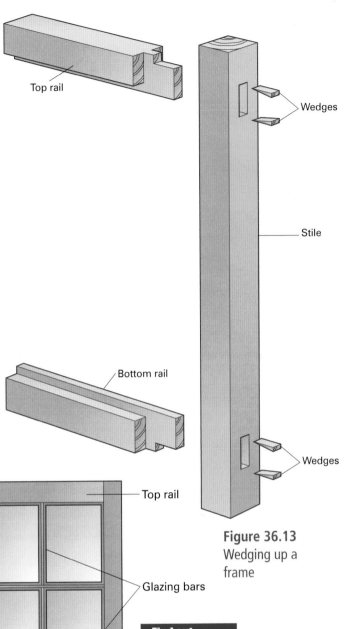

Figure 36.13 Wedging up a frame

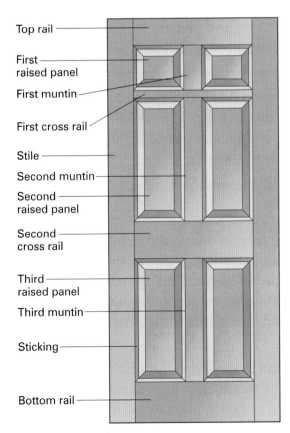

Figure 36.14 Panelled door

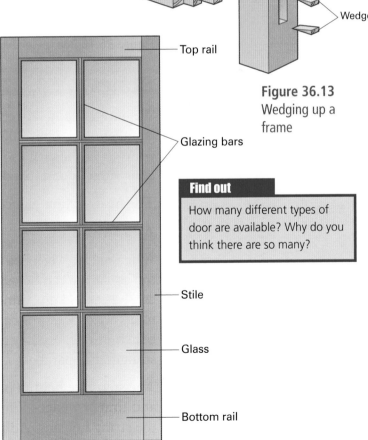

> **Find out**
>
> How many different types of door are available? Why do you think there are so many?

Figure 36.15 Glazed door

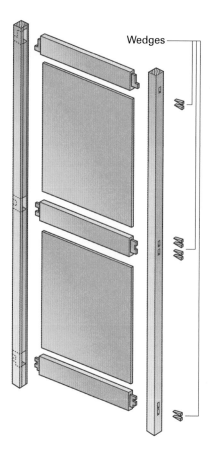

Wedges

Figure 36.16 Door assembly

General assembly procedure

The following assembly procedure can be carried out for all types of framed door.

- Assemble the frame dry to ensure that all joints are tight, the right size, square and not winding.

- Before final assembly, clean up the inside edges of all components as this will be extremely difficult once the frame has been glued.

- Glue, assemble, cramp up and check again for square and winding.

- When you are satisfied that everything is correct, the door can be wedged up.

- Clean up the rest of the frame and prepare for finishing.

Stairs

We have already looked in some detail at stairs and their installation in Unit 2008. Look at pages 158–163 to remind yourself of the terminology and regulations governing the construction and installation of stairs.

The next section will cover the assembly of stairs.

Assembly

When all the work has been carried out on the strings and newels, the assembly can be carried out. Although the assembly procedure can be carried out using a proprietary cramping system, it is more often than not carried out on an adapted workbench. The bench must be sturdy and level as it is essential that the strings are straight and parallel.

The steps are held in position in the string housing with adhesive and timber wedges. First, the treads and risers should be slid into the string housing. Adhesive is then applied to the wedges, which are then pushed into both the tread and riser housing. Wedging

> **Remember**
>
> Strings are the main boards which the treads and risers are attached to. Treads are the flat, horizontal parts of the step; risers are the vertical parts of the step. Newel posts are the uprights at the bottom of the stairs and at each turn in the staircase

must be done in a methodical order. It is easiest to start with the first riser and wedge both sides. Next, the first tread can be fitted and wedged. This process continues until all risers and treads are fitted and wedged. During this process it is important that only the wedges are glued, not the treads and risers. Finally, the risers should be screwed to the back of the tread to prevent movement when the step is being used.

Units

The most common type of unit you will probably work with is kitchen units, although if you understand the basic principles of assembling these, you will find that the assembly of other types of unit (e.g. bathroom or bedroom) can easily follow a similar procedure. Most types of unit are made from melamine-faced chipboard, medium density fibreboard (MDF), block board, plywood and, sometimes, solid timber. See Unit 2008 for more information on these materials.

There are a number of different methods of unit construction, but we will only look at probably the two most common – box and framed. We will also briefly look at different types of knock-down fittings, which are often used in unit assembly.

See Unit 2009 for further information about kitchen units and their fitting.

Box construction

This is also known as 'slab construction'. The unit is composed of vertical standards, rails and shelves and the plinth and bottom shelf are often an integral part of the unit. The back panel holds the frame square.

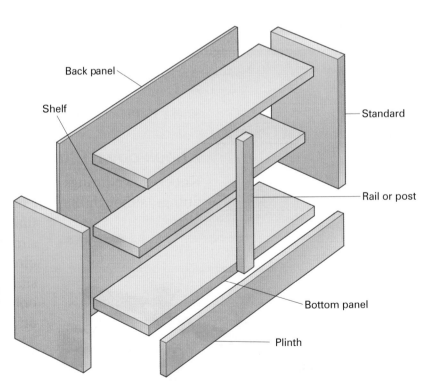

Figure 36.17 Stair assembly using a specially adapted workbench

Figure 36.18 Box construction

Framed construction

This is also known as 'skeleton construction'. The units are composed of either a pair of frames (a front and a back frame) joined together with rails, or cross-frames joined together by rails at the front and back. See Figure 36.16. The plinths and drawers are usually built separately in this type of unit. The frames are mortise and tenoned and assembly follows the same procedure as doors or casement window sashes (see pages 124–125).

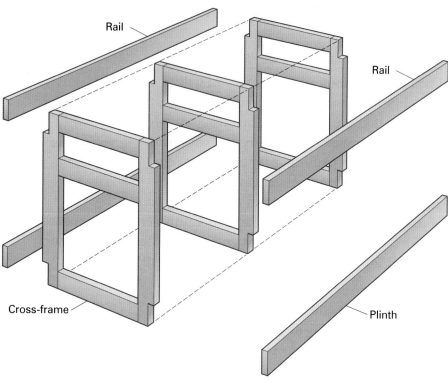

Rail

Rail

Cross-frame

Plinth

Figure 36.19 Framed construction

Knock-down fittings

Knock-down fittings were originally designed as temporary fittings, although they are often used as permanent fittings on units. The fittings are normally quite simple (a block with drilled holes) and often made from plastic. Because of their simplicity and the fact that they usually only need a few screws, they can be easily undone and the unit quickly dismantled. They are ideal for flat-pack self-assembly furniture and the joints they provide are relatively strong.

Figure 36.20 Example of knock-down fittings

Unit assembly

As with the other items, units must be dry assembled to ensure that the frame is the right size, square, and not winding. You must also remember to square up, using the procedure described on page 342, check for winding and wedge up (see page 342).

Work programme

A work programme is a method of showing very easily what work is being carried out on a building and when. The most common form of work programme is a bar chart. Used by many site agents, or supervisors, a bar chart lists the tasks that need to be done down the left side and shows a timeline across the top. A work programme is used to make sure that the relevant trade is on site at the correct time and that materials are delivered when needed. A site agent, or supervisor, can quickly tell from looking at the chart if work is keeping to schedule or falling behind.

Did you know?

The Gantt chart is named after the first man to publish it; this was Henry Gantt, an American engineer, in 1910

Working life

Assembling a door

Thomas and Laura are both second-year apprentices and have been tasked with assembling a door that has been made by an experienced joiner. They glue and cramp up the door and leave it to go to lunch. When they return, their supervisor is not happy with them as the door is out of square and the joints are not tight-fitting.

How could this have happened and what could have been done to prevent it?

FAQ

What causes a door or window frame to wind and can it be fixed?

Winding can be caused by the frame not sitting flat when it is assembled. It can also be caused by using timber with a high moisture content which warps when it dries out. Winding can only be prevented by ensuring the frame is flat when clamped up.

Why do frames always have to be dry assembled?

Every frame you make needs to be dry assembled to check that the joints are tight and that the frame is square. If a frame is not checked in this way and it is glued and clamped, and then found to have gaps in the joints, it will be difficult to put right.

Why use a double tenon joint? Would it not be easier to just put in a thicker single tenon?

Yes, it would be easier but it would not be as strong. Putting in a double tenon will increase the surface area of the joint. This will give the joint a larger area for adhesion (gluing), thus producing a stronger, well-proportioned joint.

When a stair string tenon is jointed into a newel post would you use a twin tenon or a twin tenon with a twin haunch?

Both, at the top of the stair the newel post will be deeper than the string, therefore a twin tenon should be used. At floor level the newel post will be cut flush, therefore a twin tenon with twin haunch would be the strongest joint.

Check it out

1   Prepare a method statement describing the general assembly procedure for doors.
2   Explain the difference between box and framed unit construction.
3   Explain the advantages of using knockdown fittings.
4   Explain what pitch should be used for a dovetail in softwood.
5   State the joints used in the following:
    a   stair tread to string
    b   string to newel post
    c   middle rail of a two panel door
    d   drawer on a kitchen unit.
6   Explain the purpose of glue wedges.
7   Explain what the first stage in assembly should be and why.
8   State two ways of checking a frame for square.
9   State the best way to cut out the housings on a stair string.
10   State four materials that units can be made from.

Getting ready for assessment

The information contained in this unit, as well as continued practical assignments that you will carry out in your college or training centre, will help you with preparing for both your end of unit test and the diploma multiple-choice test. It will also aid you in preparing for the work that is required for the synoptic practical assignments.

The information contained within this unit will aid you in learning how to identify and calculate the materials and equipment required to manufacture routine joinery products.

You will need to be familiar with:

- selecting the correct materials
- manufacturing joinery.

This unit will have made you familiar with the manufacture of joinery products. This builds on the setting out and marking out of joinery products covered in Units 2034 and Unit 2035. You will need to remember the information you learned in those units. For learning outcome two, this unit has introduced the correct techniques to produce components. You will need to use this knowledge during your practical assignments to be sure that you are creating the correct componets with suitable methods. As with marking out and setting out, you will need to complete work within the time allowed in the work programme and to deadlines. You will also need to be sure that you are maintaining tools correctly as the work progresses.

Before you start work on the synoptic practical test it is important that you have had sufficient practice and that you feel that you are capable of passing. It is best to have a plan of action and a work method that will help you. You will also need a copy of the required standards, any associated drawings and sufficient tools and materials. It is also wise to check your work at regular intervals. This will help you to be sure that you are working correctly and help you to avoid problems developing as you work.

Your speed at carrying out these tasks will also help you to prepare for the time limit that the synoptic practical task has. But remember, don't try to rush the job as speed will come with practice and it is important that you get the quality of workmanship right.

Always make sure that you are working safely throughout the test. Make sure you are working to all the safety requirements given throughout the test and wear all appropriate personal protective equipment. When using tools, make sure you are using them correctly and safely.

Good luck!

CHECK YOUR KNOWLEDGE

1 Which of these features is not part of a door frame?
a jambs.
b sills.
c panels.
d heads.

2 At what point in the assembly of a door or window frame is it too late to adjust any sizes?
a before gluing.
b after gluing.
c before squaring up.
d after squaring up.

3 What type of step is usually used at the bottom of a staircase?
a dog leg.
b mouse tail.
c horse hoof.
d bull nose.

4 What is the most appropriate joint to use on a kitchen drawer where the face is seen?
a housing joint.
b through dovetail.
c lapped dovetail.
d stopped housing joint.

5 One of the last things to do prior to assembly is to:
a remove the pencil lines.
b dry assemble the component.
c check for square.
d all of the above.

6 What are most units made from?
a plastic.
b melamine-faced chipboard.
c aluminium.
d all of the above.

7 How many main methods of construction are used in basic units?
a 2
b 3
c 4
d 5

8 Which of the following can be used to track work progress?
a pie chart.
b Gantt chart.
c flow diagram.
d bridle path.

9 On a window, the intermediate horizontal member between the jambs is called a:
a sill.
b muntin.
c mullion.
d transom.

10 The ironmongery fitted on an external door can consist of:
a cylinder night latch.
b spy hole.
c security chain.
d all of the above.

Index